Praise for Eli Greenbaum's *Emerald Labyrinth: A Scientist's Adventures in the Jungles of the Congo*, a *Forbes* magazine "Top 10 Science Book" of the year

"In this riveting scientific travelogue . . . Greenbaum, a professor of evolutionary genetics, recounts arduous journeys throughout the Democratic Republic of the Congo, illuminating in fluent detail its bloody history, precarious present, spectacular landscapes, and gloriously varied, now rapidly disappearing, biological abundance."—*Booklist* (starred review)

"The narrative is smooth and engaging, effectively showing the natural wonder of the Congo—and its fragility. Greenbaum's enthusiasm for his work shines through, as does his compelling message about the future of our planet."—*Kirkus Reviews*

"Greenbaum . . . combines scientific endeavor, environmentalism, and Congolese history as he shares his experiences exploring remarkable ecosystems in the midst of a civil war."—*Publishers Weekly*

"Greenbaum's account of a 2008 expedition with Congolese colleague Chifundera Kusamba and a crack team of local rangers is much more than derring-do among prodigious natural riches: it is also a meditation on how colonial power seeds violence. A valuable record of conflict and conservation at a time of climate change and population pressures."—*Nature*

"Delivers the goods. . . . Greenbaum masterfully wraps each of his biological discoveries in rich tapestries, from Africa's history to its current quagmires of politics and corruption."—Kurt Johnson, coauthor of *Nabokov's Blues: The Scientific Odyssey of a Literary Genius* and *Fine Lines: Vladimir Nabokov's Scientific Art*

"Engrossing. . . . Slipping unobtrusively between armed militias, avaricious bureaucrats, courageous conservationists, and tropical diseases, Greenbaum takes his readers with him to discover the depths of our ignorance of Congo's natural history, with erudite forays into the country's political and social history and its importance to the fate of the planet. . . . His research is urgent."—Ian Redmond, OBE, wildlife biologist, and ambassador for the United Nations Convention on the Conservation of Migratory Species of Wild Animals

"Thanks to scientists like Greenbaum, expeditions such as those described in this book offer reassurance that innovative research can proceed despite the many difficulties that get in the way."—Jonathan Kingdon, University of Oxford

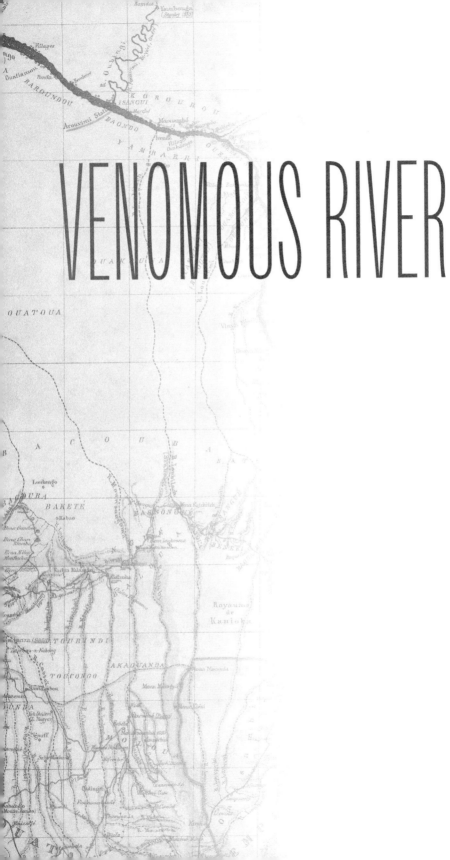

VENOMOUS RIVER

VENOMOUS

Eli Greenbaum

HIGH ROAD BOOKS

ALBUQUERQUE

RIVER

Changing Climate,

Imperiled Forests,

and a Scientist's Race

to Find New Species in the Congo

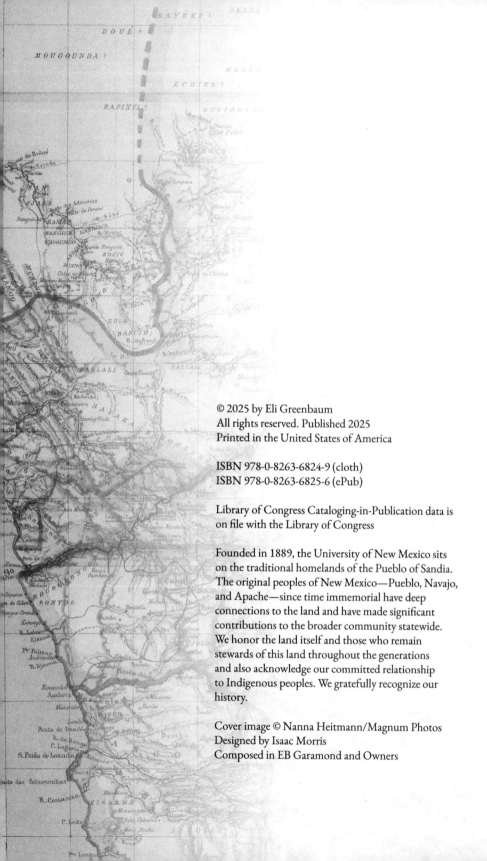

ISBN 978-0-8263-6824-9 (cloth)
ISBN 978-0-8263-6825-6 (ePub)

Library of Congress Cataloging-in-Publication data is
on file with the Library of Congress

Founded in 1889, the University of New Mexico sits
on the traditional homelands of the Pueblo of Sandia.
The original peoples of New Mexico—Pueblo, Navajo,
and Apache—since time immemorial have deep
connections to the land and have made significant
contributions to the broader community statewide.
We honor the land itself and those who remain
stewards of this land throughout the generations
and also acknowledge our committed relationship
to Indigenous peoples. We gratefully recognize our
history.

Cover image © Nanna Heitmann/Magnum Photos
Designed by Isaac Morris
Composed in EB Garamond and Owners

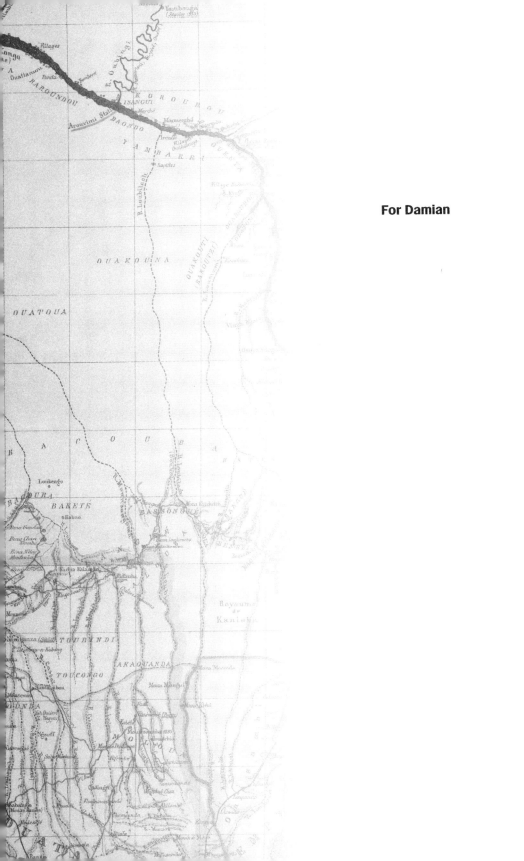

For Damian

CONTENTS

Contents

PREFACE

In 1899 when Joseph Conrad published his famous novella *Heart of Darkness*, he described the Congo River as "a great snake." As a herpetologist, evolutionary biologist, and seasoned expedition leader 114 years later, I wanted to find Congo's bona fide serpents,[1] along with the country's rarest frogs, lizards, crocodiles, and turtles, which live in and along Africa's second-largest river.[2] I would not be disappointed by the incredible biodiversity of Congo's animal fauna, and DNA-based analyses of the natural history specimens I would later analyze in my lab confirmed that many of the species I had encountered were completely new to science. From the standpoint of a biologist, an expedition into the heart of Congo is akin to walking on the moon, because one does not know what incredible species, known or unknown, are lurking around the next bend in the river.

Although tropical forests cover less than 10 percent of Earth's land surface, they are home to around two-thirds of the planet's biodiversity.[3] The vast majority of this biodiversity remains completely unknown.[4] Compared with the widespread deforestation that has occurred in recent decades on an industrial scale in the Amazon and Asian tropics, the Congo Basin of Central Africa, the world's second largest tract of rainforest, has suffered relatively moderate damage, mainly due to its remoteness, lack of infrastructure, low population density, prohibitively high risks for investors, macabre smorgasbord of deadly tropical diseases, and stubbornly resilient political instability.[5] These factors have also discouraged scientific exploration of these rainforests and explain how it is likely to be—in the context of biodiversity—the most poorly known terrestrial region in the world.[6]

Unfortunately, the world's forests are being assaulted by a variety of human-induced afflictions, including, just to name a few, global climate change effects, including increased carbon dioxide levels in the atmosphere (leading to disturbances in plant ecological interactions), freshwater degradation, unsustainable exploitation of forest species and resources (including fossil fuels, logging, and mining in the Congo Basin), introduction of harmful invasive species, eutrophication from fertilizer runoff, toxic pesticide use, and most damaging of all, widespread destruction of the forest itself from road construction, firewood gathering, and conversion of the land to farms and pastures from a dramatically increasing human population.[7] Much of the world does not yet realize that this devastation is an existential threat, because the human species cannot survive without the myriad "ecosystem services" that these forests provide across the globe, including the sequestration of massive amounts of carbon dioxide in our atmosphere to stave off the potentially lethal effects of global climate change.[8] Climate change has become impossible to ignore, and yet, many of us are going about our daily lives only vaguely aware that we are losing our only chance to avert disaster now.

Conservation of the remaining natural areas of the Congo Basin will be absolutely crucial to avoid a climate catastrophe in the coming years, and its undiscovered biodiversity undoubtedly includes new species with secretions and toxins that will be of enormous benefit to humanity, including antibiotics, antivirals, analgesics, and a cornucopia of other bioactive molecules that could be developed into drugs to treat everything from cancer to high blood pressure.[9] Approximately 25 percent of medicines that have been patented by the Western world were first identified from rainforest plants that were in use by Indigenous people.[10] Of 185 cancer treatment drugs developed since 1981, 65 percent were derived or inspired from natural plant or fungi products, some of which are now threatened with extinction.[11] This book discusses several examples of venomous reptile venom components that have been modified into miraculous drugs like Ozempic, but skin secretions of frogs contain many potentially life-saving peptides too. To give just one example, the so-called Yodha (Sanskrit for *warrior*) protein that was isolated from a frog in India is "virucidal" against Zika and dengue viruses.[12]

Because many species of amphibians and reptiles are likely being pushed to extinction before the scientific world is even aware of their existence, it is imperative to conduct biodiversity surveys while there is still time. The year before I undertook the expedition described in this book, I was awarded a grant (DEB-1145459) from the US National Science Foundation (along with colleague Kate Jackson) to study the amphibians and reptiles of the Congo Basin, and many of the results I present in this book, obtained in the decade since the expedition occurred, are a result of that grant. My scientific career in Congo would have been impossible without the dedicated work and support of several Congolese colleagues and institutions, and I am proud to say that our nearly two-decade-long scientific collaboration has led to scores of peer-reviewed scientific publications, including many descriptions of spectacular new species. The information we have published has helped to create enthusiasm for improvements to protected areas in Congo, an essential endeavor to mitigate losses from the ongoing extinction crisis, which is hitting amphibians harder than any other major group of vertebrates.[13]

As this book was nearing completion in January 2025, the Congo government miraculously announced plans for the Kivu-Kinshasa Green Corridor, which will become the largest tropical forest reserve in the world. Based on the preliminary map of the reserve that was released with the announcement, it seems that nearly all of the wild places I describe in this book will become part of this reserve.[14] As you can imagine, establishing such a huge protected area (about the size of France) will be an enormous challenge,[15] but we should celebrate this good news and cheer on the Congolese people.

It is my sincere hope that in this book, the reader will learn about the importance of biodiversity surveys and natural history collections, humanity's sometimes invisible links to this biodiversity and the environments in which they are concentrated (especially tropical forests), and the dire need to conserve as much of these natural environments as possible. We will need them to survive the climate change crisis that is becoming blatantly obvious as the planet gets hotter every year.

As I noted in the preface to my previous book *Emerald Labyrinth*, "As a white man working in Africa, and considering the shameful history of other white men in Africa, I fear it is impossible to convince every reader of this book that I have made every effort to be fair to the amazing Congolese people I have worked with over the years, but as imperfect as my efforts may have been, that was my goal. It is my sincere hope that my actions have improved the lives of the people (and their families) I have worked with, and that my work will be a great benefit to the country I have been privileged to know and love."[16] The narrative you are about to read is true, but to safeguard my team on future expeditions and protect the privacy of some people we encountered, some minor details have been modified.

Eli Greenbaum

March 2025, Las Cruces, New Mexico

LIST OF ABBREVIATIONS

acute coronary syndrome: **ACS**

Alliance of Democratic Forces for the Liberation of Congo: **AFDL**

angiotensin-converting enzyme: **ACE**

acquired immunodeficiency syndrome: **AIDS**

Avtomat Kalashnikov assault rifle: **AK-47**

Before Current Era: **BCE**

Broadband Global Area Network: **BGAN**

Centre de Recherche en Sciences Naturelles: **CRSN**

Convention on International Trade of Endangered Species of Wild Fauna and Flora: **CITES**

Current Era (i.e., Anno Domini): **CE**

deoxyribonucleic acid: **DNA**

Direction Générale de Migration: **DGM**

endogenous viral elements: **EVEs**

Food and Drug Administration: **FDA**

Global Amphibian Assessment: **GAA**

global positioning system: **GPS**

Global Reptile Assessment: **GRA**

human African trypanosomiasis: **HAT**

human immunodeficiency virus: **HIV**

human papillomavirus: **HPV**

Institut Congolais pour la Conservation de la Nature: **ICCN**

Institute for Scientific Research in Central Africa: **IRSAC**

International Union for the Conservation of Nature: **IUCN**

Kuramo Capital Management: **KCM**

light-emitting diode: **LED**

million years ago: **mya**

National Institute of Health (USA): **NIH**

National Science Foundation (USA): **NSF**

non-governmental organization: **NGO**

parts per million: **ppm**

Plantations et Huileries du Congo (Congo Plantations and Oil Mills):
PHC

ribonucleic acid: **RNA**

Snake Detection Theory: **SDT**

Species Survival Commission: **SSC**

temperature-dependent sex determination: **TSD**

United Nations: **UN**

United States of America: **US**

University of Texas at El Paso: **UTEP**

World Health Organization: **WHO**

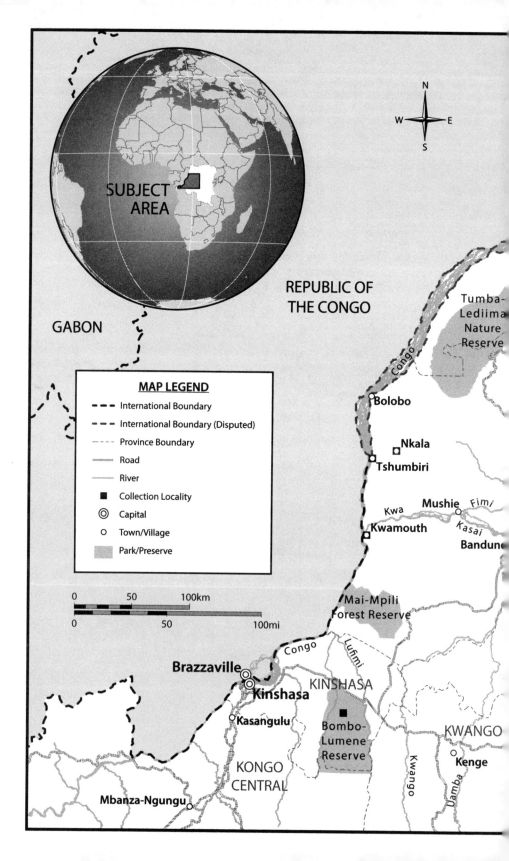

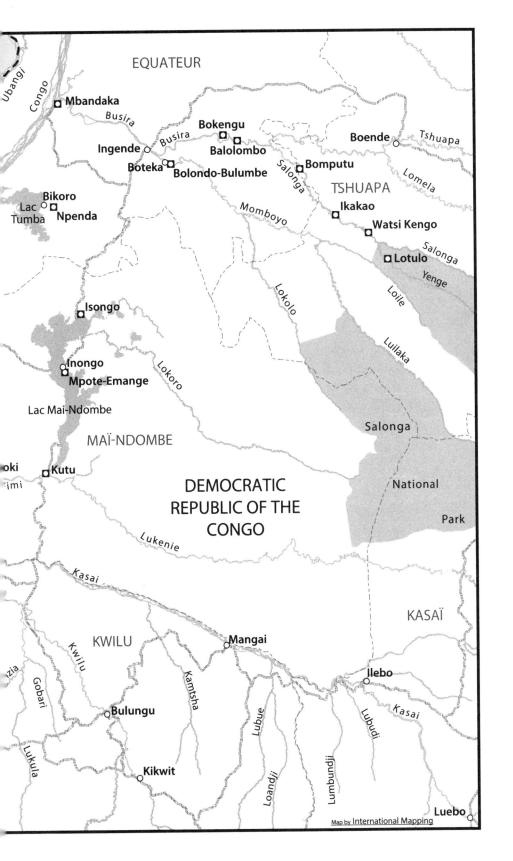

DRAGON'S BREATH 1

Beneath the veil of a mosquito head net, my bloodshot eyes cracked open only slightly to perceive the gloom of dawn transitioning into a haze from the sun's passage over a drizzling sky. The voracious appetites of the mosquitoes that had tormented me through the night had mostly ceased, and I closed my eyes again in a hopeless attempt to rest a little more, just as the rickety boat slowly came to life. An old man groaned as he strained to free his stiff body from the starboard bulwark, the same nail-studded frame I had used for my own pillow. A young woman emitted a quiet grunt before jerking awake with a start, suffering a nightmare, no doubt, from some nasty memory in the past. Two men whispered to each other in Lingala, the lingua franca of western Democratic Republic of the Congo (Congo for short), trying to mute their conversation out of politeness to the sleeping passengers. But they barely managed to hear each other over the tranquil cascade of raindrops that were plopping all around us and into the Congo River.

I knew it was just a matter of time before the other twenty passengers would start conversing, and resigned to my exhausting fate, I opened my filmy eyelids, this time for good. As I slipped off the raincoat that had served as my blanket, every muscle of my neck seemed to be twisted into Gordian knots. With a quiet grunt, I pushed my body away from the slimy flank of the boat, and stretched my stiff legs out as far as I could without disturbing the people who were still sleeping around me. My feet swished through several inches of water the color of strong tea, which had accumulated in the bottom of the boat after the crew had abandoned their relentless efforts to bail it out in the middle of the night. The hours-long journey the day before, coupled with a bitterly cold night on the edge of the river, had taken its toll on me, both physically and mentally. My mind drifted briefly to the happy memories from the week before, when I had indulged in the luxury of swanky hotels in Bali, only to rush off forty-eight hours after my return to Texas to catch a plane to Africa. Completely unprepared for the draconian conditions in which I now found myself, my body mutinied by launching wrenching aches from every hidden crook.

A bewildering throbbing sensation commenced in my skull, reminding me that my addiction to coffee had not been satiated. Certainly, it would be hours before somebody would be able to take the time to boil some of my purified water, especially since our cook Elvis had taken one look at our leaking vessel the previous day in Kinshasa and promptly quit on the spot. How could an eighteen-year-old kid, whimsically named for a flamboyant rock star, have had more sense than I did? I glimpsed the pocket of my backpack, knowing that it contained a couple of precious Via packets of Starbucks instant coffee. But if I took one out now, I would have to mix it with the tainted remains of a bottle of water I had bought in the capital the day before, leaving me nothing to drink for the day-long journey that remained. I watched as some of my boatmates brushed their teeth with water directly from the Congo and pondered what the river water might do to me if I got desperate enough to drink it.

Slowly, my eyes adjusted to the light outside the dim shade of the canopy, and I realized that the boat was enveloped in mist. My weary mind, claustrophobic from the unexpected oppression of thick air, invented a bestiary of creatures just beyond my sight. I pondered whether it was possible for a dragon to have spewed out the wispy clouds of fog. After all, my dreamlike brain reasoned, enormous dinosaurs roamed across Africa millions of years ago. *Carcharodontosaurus* was especially terrifying—named for the resemblance of its eight-inch-long teeth to those of the great white shark, these carnivorous theropod dinosaurs were comparable in size to their more famous American cousin *Tyrannosaurus rex*.[1] Was it *really* beyond the realm of possibility that an ancient lineage of monsters had survived in the remote obscurity of Congo? In neighboring Republic of the Congo, Lake Télé had its own version of the Loch Ness Monster, known as Mokélé-mbembé.[2]

I began to perk up a little as my imagination invented spectacular new species of creatures that had somehow escaped scientific notice before I came to search for them. Few biologists, and even fewer herpetologists, had taken the risk to search the deepest forests in western Congo for their troves of poorly known biodiversity, and indeed, this is the reason I had come. But even if my wildest dreams could come true and I encountered the reptilian equivalent of a unicorn, would I succumb to some nasty disease before I could return with my scientific discoveries? To fight off contagion from what is arguably the most well-endowed morass of tropical disease in the world, including malaria, sleeping sickness, typhoid fever, and God forbid, Ebola, I would need a healthy constitution, adequate rest, clean water, and good nutrition. In other words, I would need a miracle. As the first rays of sunlight began to burn the mist away, reality set in, and I felt hopelessly foolish, thirsty, hungry, and demoralized.

Close to the bow, a ramrod body sat up in a full-length woman's winter coat. I recognized him as the striking man we had met the day before as we pulled away from the bank at Kinshasa. Avoiding the port authorities that had delayed us for hours, he had jumped onto the small boat at the very last minute. Speaking in Mashi, the native language of my colleague and logistics coordinator Chifundera Kusamba, he had convinced all of us that he could replace Elvis as our expedition cook for the next ten weeks. With a sarcastic smile, he told us his name was Poignard, French for dagger. Unfazed by the bullshit artist, we laughed at his exaggerated swagger and charming stories, and having no alternative, I agreed to hire him on the spot.

But now as I regarded him against the shadowy backdrop of the retreating haze, his eyes were obscured by the gloom and his face was expressionless. A preternatural voice seemed to whisper that he was staring at me. I felt slightly uneasy without knowing why. For a moment, I thought of Central Africa's brutal wars, which unleashed legions of murderers who killed hundreds of thousands and answered to no one for their horrific crimes. Years after the fires of battle have disappeared, warzones continue to cast long shadows. But the human chameleon in my presence had been adapting to his surroundings for two decades, and dismissing my paranoia, I returned my attention to the tendrils of mist as they crept along the river's dark surface. Even then, I should have known that there are far more dangerous things in Congo than dragons.

Two weeks before my memorable boat ride on the Congo, and many hundreds of miles to the east in the ruggedly beautiful Albertine Rift Mountains, my colleague Wandege Muninga[3] captured a very dangerous snake. The roughly foot-and-a-half long (53 cm), grayish-black serpent had emerged from its subterranean haunt after heavy rains flooded it out, and Wandege spotted it just after dusk on his way to visit a friend in the small town of Lwiro. Being an experienced snake hunter, and armed with foot-long metal forceps (think an oversized pair of tweezers), he safely picked up the variable burrowing asp (*Atractaspis irregularis*) and wrangled it into a cloth bag that he always kept with him for incidental snake captures. Just in case, he also carried a supply of the *katumbanyi* plant, which is supposed to have anti-venomous properties. This *Alchornea* plant in the family Euphorbiaceae[4] was in use by forest-dwelling Indigenous people near Mwenga when Chifundera learned about it in 1986.[5] Since then, every "snake man" at Centre de Recherche en Sciences Naturelles (CRSN), including Wandege, has never left home without it.

One could argue that this snake was designed by the devil, and Wandege was wise to capture the asp, otherwise known as a stiletto snake, with extreme caution. Evolving in the Oligocene about twenty-seven million years ago, the common ancestor of *Atractaspis* snakes pursued a novel strategy to hunt nestling rodents and reptiles in their burrows.[6] The cramped spaces of the subterranean tunnels prevented the venomous[7] snakes from biting their prey with forward strikes in a similar manner to vipers and cobras. Out of necessity, evolution invented a unique mechanism that allowed the snakes to use their unusually long fangs to side-stab their prey via a short, rotatable maxilla bone, the same one that contains your upper teeth. The deadly invention allowed the snakes to rapidly dispatch an entire family of rodents without opening their mouths, simply by protruding their long, wicked fangs in a manner akin to a stiletto knife. After confirming death with fang pricks, they use one fang as a gaff to drag the prey out of the burrows and consume them.[8]

To prevent a counterattack by toothy mammalian prey, the venom is required to work quickly, and over millions of years, evolution's alchemy lab designed new, fast-acting toxins to ensure the snakes could be efficient killers. Because humans have a similar physiology and cell structure to rodents, some stiletto snake venoms can cause terrible damage to snakebite victims. Depending on the species of stiletto snake, bites to humans can range in severity from temporary pain and swelling to a dramatic rise in blood pressure, tissue necrosis (blackened, gangrenous fingers that rot away or lead to amputation), respiratory failure, heart problems, cerebral hemorrhage, and even death as quickly as forty-five minutes.[9] In one infamous case study, a hapless, heavy sleeper died from eight stiletto snake bites after the serpent wandered into his bed in the dead of night.[10] One can only imagine the ghastly nightmare that surely raged during his final thoughts.

The species captured by Wandege happened to be the same one responsible for the deadly slumber, and he certainly understood the importance of securing it, because snakes are born escape artists. Soon afterward, I was told by Chifundera, he traveled south with his dangerous captive to Bukavu. From there, he took the day-long ferry over Lake Kivu to the city of Goma, near the northwestern border of Rwanda. His plan was to rendezvous with other Congolese colleagues who had worked with me during previous expeditions, including Chifundera, the logistics coordinator and lead Congolese herpetologist. Shortly after dusk, Wandege decided to take a nap, and he awoke to screams from a nearby woman who had noticed that his snake escaped through a tiny hole in the bag. He quickly realized that if the panic spread on the overloaded boat, it could capsize, and to avert disaster, he recaptured the snake with his hand. Big mistake.

Although he managed to calm the woman and secure the snake in a bucket that was within reach, it bit him during the chaotic scramble to recapture it. Knowing that he was in trouble, he grabbed a razor blade and slashed deep cuts in his flesh near the fang puncture, and then stuffed them with his *katumbanyi* plant. Perhaps it worked a little, because he was able to find his forceps to secure the stiletto snake in a new bag. He also managed to use his cellphone to call his son in Goma, and asked him to meet him at the dock with crushed leaves of the *Lantana trifolia* plant (another one with putative anti-venomous properties), commonly used across the world as an ornamental hedge because of its pretty lavender flowers.

Wandege planned to combine the leaves with water to make an oral elixir. But by the time he arrived in Goma the next morning, his entire arm was painfully swollen and he was taken straight to the hospital. There he languished for hours in agony, and at one point during the night, the doctor warned Chifundera that Wandege was going to die.

But Wandege's will to live was strong, and somehow he managed to pull through. When I found out about the incident the day before leaving the United States, I told Chifundera that Wandege could keep the advance I sent him for the expedition, and that he should really go back to Lwiro to recover. But he recovered quickly and refused to return home. As a fellow snake enthusiast, I understood his reluctance to miss the expedition. Where we were headed, scores of colorful species of snakes would surely be hiding around every corner, like coiled Christmas presents waiting to be unwrapped from the jungle. Despite knowing painfully well the price he might have to pay, how could he resist? None of us could.

With Wandege's participation confirmed, we planned to meet in Kinshasa, Congo's riverside capital. Full of the hope one experiences at the beginning of an adventure, I looked forward to a cold Primus, effectively the national beer of Congo, while enjoying one of the delicious grilled fish from the nearby river. Although I had been to eastern Congo six times before during previous expeditions, this was to be my first trip to the western part of the country, over 1,400 miles away. Based on travel books, I knew the megalopolis would be hot, polluted, chaotic, and dangerous in places, a far cry from the "Kin la belle" (Kinshasa the beautiful) it had been called during the colonial era. But Congo's largest city, now free from the injustice of colonialism, also promised to be an interesting crossroads of the country's cultures, politics, and history.

Long before any white man explored the lower Congo River where Kinshasa is located today, much of the territory surrounding it had been part of the kingdom of the Kongo, which included millions of people and extended far to the south into modern-day Angola.[11] The kingdom's rise likely coincided with the introduction of blacksmithing around the fourteenth century, and indeed, one of the Kongo's first kings, Ntinu Wene,[12] was widely known as Ngangula a Kongo (Blacksmith [King] of the Kongo) for his interest in forging iron weapons and tools. Metalworking was usually restricted to nobility, and it became an honored part of several rituals, as well as an essential component of the kingdom's art, culture, and violent expansion.[13]

Kongo civilization was remarkable in many ways, and although specific practices varied among ethnic groups,[14] some examples paint a vivid picture of the society. Its judicial system was run by a chief justice and several regional judges, who laid down the law, enforced it, and settled disputes. Diviners were also influential members of Kongo society who determined causes of death and illness (usually from displeased ancestors, evil fetishes, or sorcery), advised nobles about optimal times for war, interpreted omens, and administered ordeals by poison to ascertain the guilt or innocence of accused thieves, sorcerers, and "eaters of souls."[15] A ritual specialist called *nganga* hammered nails or blades into humanlike wooden power objects that were called idols or fetishes by European observers.[16]

Bakongo (the people of Kongo) cultivated raffia palms, from which they were able to make astonishingly fine fabrics that some Europeans compared to velvet and silk, and this cloth was used as currency. A Kongo week was four days long, including one day (*nsona*) dedicated solely to religious rites and recognition of ancestors.[17] Because of their sensitive perceptions, dogs were thought to bridge the worlds of the living and dead ancestors.[18]

Like other notable civilizations of Africa, Kongo had an impressive military. One European observer in 1595 estimated that the king had 100,000 men in his service, and when they went to war, officers used sophisticated signals to coordinate attacks with men using bows and arrows. Armed with daggers and shields made of bark or buffalo hide, soldiers wore small bells on their belts to give them courage during fighting, and the officers issued commands via drums, bells, and horns made from ivory, creating a dramatic soundtrack to their battles. Among the desired spoils of war were human beings, who were either kept as slaves by the most successful soldiers, or more commonly, sold to Portuguese, Dutch, and other European slavers. Over time, the slave trade corrupted the Bakongo to such a degree that even nobles were sold into slavery, weakening the traditions and institutions that had formed the nucleus of the kingdom. The end came at the Battle of Ambouila (a.k.a. Mbwila) in present-day Angola in October 1665, when the Portuguese defeated the Kongo army and decapitated their king.[19] The kingdom had been effectively destroyed, but as I would observe firsthand, some aspects of its language and culture would endure.[20]

Although most of the roughly 2,900 miles of the Congo River is navigable, a series of 32 cataracts begins at Matadi, 93 miles east of the river's mouth at the Atlantic Ocean, and this obstacle prevented inland exploration by Europeans for centuries.[21] Henry Morton Stanley, a British explorer, led a grueling 999-day expedition from Zanzibar to Lake Tanganyika, and thence down the Congo River to Matadi in 1877, losing all four of his European companions and scores of African escorts along the way to disease, drowning, and relentless skirmishes with cannibal tribes who screamed for "meat!" during their attacks. As Stanley neared the end of this journey at the upper reaches of the cataracts, he noticed an unusual widening of the river, which his British companion Frank Pocock supposedly suggested should be named Stanley Pool.[22] Stanley was the first white explorer[23] to mention a "populous settlement" on the southern edge of the pool named "Nshasa," where he traded for some badly needed food from its king.[24] After encountering some Bakongo traders lower down the river at Mpakambendi, he was struck by their relatively small stature and "good looks."[25]

Stanley's exploration of the river led to the establishment of the Congo Free State in 1885, eventually with Léopoldville as its capital (named after King Leopold II of Belgium who personally owned and ruthlessly exploited this state), complete with a railroad to circumvent the roughly two hundred miles of cataracts between Matadi and the capital, which was initially established as a trading center and located at the southern edge of Stanley Pool (now known as

Malebo Pool).[26] In 1896, as the railroad was nearing completion, a French rubber trader described Léopoldville as "an attractive and well-built town" with straight avenues that passed through plantations of coffee, bananas, and other tropical fruits.[27] At that time, Kinshasa seems to have been considered a nearby, separate town where steamships were assembled from piecemeal, sixty-pound sections that were brought by Bakongo porters from Matadi.[28] As the twentieth century dawned and continued, Léopoldville grew in population and importance.

By 1956, at the height of the colonial era in the Belgian Congo, Léopoldville resembled a modern, metropolitan city in Europe. A Belgian tourist guide counted 16,355 Europeans and 282,000 "Natives" that lived in four main areas of the city, including Kinshasa (now merged with the capital city). The city contained several medical marvels, including hospitals (separate ones for Europeans and Congolese in line with colonial modus operandi), a vaccine institute, dispensaries to treat sleeping sickness, tuberculosis and venereal disease, and numerous chemical and biomedical laboratories.[29] Léopoldville also contained a radio and telegraph network, several colleges and schools, headquarters for scores of colonial businesses, numerous churches and religious institutions, and sixty-five thousand spectators could watch soccer games at King Baudouin Stadium.[30]

A decade later in the hasty and violent aftermath of the collapse of Belgian colonialism, and with the assistance of the US government, a military commander named Joseph-Désiré Mobutu (a.k.a. Mobutu Sese Seko) seized power in a coup and became dictator of the country. Many of the colonial institutions and their associated infrastructures decayed from neglect in the subsequent three decades of Mobutu's rule. He renamed the country Zaïre, flipped Léopoldville to Kinshasa, and continued to use the renamed 20 May Stadium for soccer games, but it also served as an infamous dungeon for torture and execution of political opponents.[31] In 1974, Mobutu hosted the famous "Rumble in the Jungle" fight between George Foreman and Muhammad Ali, attended by sixty thousand people in the stadium.[32] Because of some disastrous economic decisions by Mobutu in the 1970s, as well as his kleptocratic regime, Zaïre fell on hard financial times.[33] Millions of Congolese seeking a better life swelled into Kinshasa's slums until the population approached about 5 million at the end of the twentieth century.[34] By

2015, the megacity's population had doubled to about 10.6 million, and current projections predict it will be a whopping 75 million within fifty years[35]—more people than the entire population of the country in 2014.[36]

Following the brutal reverberations of the Rwandan genocide in 1994, Mobutu's regime imploded in 1997 when a Congolese rebel leader named Laurent-Désiré Kabila marched hundreds of miles from eastern Congo to Kinshasa, committing numerous atrocities with his *kadogo* (meaning "small one" in Swahili) child soldiers along the way. Kabila installed himself as Mobutu's replacement, reverted Zaïre to Democratic Republic of the Congo, and in 1998 when he turned on his foreign backers from Rwanda and Uganda, the Second Congo War (a.k.a. Africa's World War) commenced, eventually leading to the deaths of over five million people. Kabila was assassinated by one of his own bodyguards during the conflict in 2001, and power was transferred to his son Joseph Kabila, who ruled Congo as president from the historic Palais de la Nation in Kinshasa from 2001 to 2019.[37]

After the exhaustion of twenty hours of flying time from El Paso to Atlanta, Paris, and Kinshasa, with hours of layovers in between, my arrival in Congo was a blur. Emerging from the cold dry cocoon of the plane on to the tarmac in the early evening, a blast of humidity seemed to instantly bathe my body in dew. As I waited with my fellow passengers for a bus to transfer us to the arrivals hall, squeaking noises alerted me to the presence of numerous bats above me, undoubtedly taking advantage of the insect life attracted by the airport's bright lights. According to Professor Guy-Crispin Gembu Tungaluna, a bat specialist at the University of Kisangani, Congo (and neighboring Rwanda and Burundi) has over 130 known species of bats, and like many other vertebrate groups, more are being discovered on a regular basis.[38]

While waiting in line at the immigration checkpoint, I was only vaguely aware of the mosquitoes helicoptering under the yellowish haze of fluorescent lights, and I contemplated the sober reality that nothing would be truly comfortable again for months. Then my attention was riveted by a large color poster showcasing the best of Congo's wildlife, including the chimpanzee, lion, elephant, okapi, and rhino. My eyes lingered on the latter animal, and I could not help feeling saddened by it. Most of the tourists probably did not realize that Congo's native rhinos are long gone.

On the one hand, the black rhinoceros (*Diceros bicornis*) was wiped out of most of north-central Africa, including eastern Congo, long ago.[39] In previous centuries, some of this decline was caused by trophy hunters—in 1909 former US president Theodore Roosevelt and his son Kermit killed eleven black rhinos during a safari to East Africa.[40] In more recent decades, demand from Asian markets has fueled devastating African rhino poaching because of the bogus belief that rhino horn (consisting of keratin protein like that in your fingernails) can be used to cure everything from cancer to hangovers and erectile dysfunction.[41] Today there are about three thousand mature individuals remaining in southern and eastern Africa. The International Union for the Conservation of Nature (IUCN), the gold standard for threat assessments of our planet's life, lists the species as critically endangered, the most serious threat level before extinction in the wild.[42] However, thanks to extremely dedicated anti-poaching units in many protected areas and private game reserves of Africa, including the female "black mambas" of South Africa,[43] black rhino populations are increasing,[44] and recently, they were reintroduced to parks where they had been absent for several decades.[45]

More disheartening is the annihilation of the northern white rhino (*Ceratotherium simum cottoni*), a distinct subspecies from the more common and widespread southern white rhino (*C. simum simum*).[46] A genetic study of these rhino subspecies from 2016 suggested they diverged from each other less than one million years ago with only 0.009 p-distance (i.e., the proportion [*p*] of DNA nucleotide sites that are different between two compared sequences[47]), not nearly enough genetic distance to be considered distinct species, but nonetheless, they were supported as unique evolutionary lineages, each in need of conservation.[48] The northern white rhino had a historical distribution in open savannas with scattered combretum trees in an area encompassing northwestern Uganda, northeastern Congo, South Sudan, and adjoining Central African Republic and southern

Chad.[49] In the 1960s, their numbers had been at least two thousand.[50] When the First Congo War reached Garamba National Park in northeastern Congo in 1996, the only place in the world where the remaining thirty-one rhinos lived in the wild at that time, another poaching spree of the park's large mammals commenced. By 2006, only a single individual was spotted during a thorough aerial and ground search of the park,[51] and 2007 was the last year a white rhino was seen in the wild.[52] By 2018, only three of the rhinos remained in captivity, all transferred from a zoo in the Czech Republic to the Ol Pejeta Conservancy in Kenya, where experts hoped the natural environment would encourage reproduction.[53] Sudan, the last surviving male and only longshot hope for a captive breeding program, died that year, his final moments memorialized in a heartbreaking photo that made the cover of *National Geographic's* October 2019 issue.[54]

Is there any silver lining? Yes, but patience, optimism, and imagination are required. Rhino sperm, eggs, and other cells in frozen tissue banks might be used with cutting-edge techniques in reproductive biology to implant lab-fertilized embryos into surrogates, but many technical challenges need to be surmounted.[55] In recent years, scientists have produced twenty-nine northern white rhino embryos from one of the remaining female's eggs and sperm from four deceased males. In 2024, a technique for rhino in vitro fertilization was perfected, and there is new hope to implant a northern white rhino embryo in a southern white rhino surrogate.[56] However, even if this promising endeavor is successful, the genetic diversity of the babies from only one mother will be low, which is likely to cause problems with inbreeding depression—lower survival and fertility rates of progeny from closely related individuals.[57] Additional innovations from stem cells and gene-editing technology will be needed to address the latter problem, all of which are likely to require decades of work.[58] Recently developed advances in genomics sequencing have made it possible to obtain impressive amounts of genetic data from historical natural history collections of rhino skins and horns obtained as early as 1845,[59] and together with the revolutionary CRISPR-Cas9 gene-editing tool, this, too, could lead to a path to restore the northern white rhino to the wild in the coming decades.[60]

As for the southern white rhino, the subspecies had an estimated population size of about ten thousand individuals in 2020, but poaching continues to be a problem in areas with weak law enforcement.[61] In June 2023, a conservation organization called African Parks transported sixteen southern white rhinos from a private reserve in South Africa to Garamba in an attempt to "fulfill the role of the northern white rhino in the landscape."[62]

An annoyed immigration official jolted me out of my thoughts and jammed his stamper onto my passport with the force of a sledgehammer before waving me through to the arrivals area. Restless and struggling with the sauna-like climate, I wandered into the chaos near the luggage conveyor belt, where a sweaty man with a dolly greeted me in French, asking if I needed help with my baggage. Through the throng of people near the exit, I could see Chifundera waiting eagerly for me, but the airport official holding them back looked serious and would not let anyone pass beyond a flimsy security barrier. Knowing that it would be difficult to collect the mountain of luggage I had brought with me and push it through the tight crowds of people by myself, I reluctantly agreed to accept some help. I fidgeted nervously as I watched and waited for an hour for my luggage to appear, but by some miracle, nothing was missing or damaged.

"Oh!" Chifundera exclaimed as I reached out my hand to shake his. "The food in Texas must be very good!" he joked as he eyed my waistline, undoubtedly larger from the last time we had seen each other. His warm smile was very welcoming, and I caught a glint of excitement in his eye, because our seventh expedition together was about to begin. He started barking orders at the baggage handlers wrangling my stuff, and when somebody on the airport security staff started wondering out loud what I might have in all my baggage, Chifundera immediately switched his expression to aggrievement, going out of his way to tilt his head and expose an evil eye toward the offender. Somehow he looked younger than his diminutive middle-aged body suggested, but ever the politician, his natural charisma and regal composure persuaded the airport officials to let us pass through the barrier without delay. This would be only the first of many bureaucratic miracles he would conjure during the expedition.

As we emerged out of the arrivals area, I spied two beaming faces from Aristote and Wandege, the other veteran members of the Congolese herpetology crew. Stocky and of medium height, Aristote gripped my extended hand and drew me in close so that he could greet me in the traditional African way with a strong tap of our shaved heads, first on the left, then on the right, and finishing at the center. His enthusiasm (or was it the beer that I could smell on his breath?) was so strong that his final tap was more of an accidental headbutt, but I forgot my headache the instant I saw his huge grin. "Jambo, muzungu (Welcome, white man)," he slurred as he shook my hand repeatedly. Invariably jovial, Aristote never failed to put a smile on my face.

More cautious and quiet, a tall and skinny Wandege slipped in to repeat Aristote's greeting, exclaiming "Welcome!" in his thick and abstruse accent. I looked him up and down as well as I could in the dim light to see if I could sense any aftereffects of the snakebite, but he looked completely normal. When I asked him how he was feeling, he replied "Ok, no parabolem!" Before I had a chance to contemplate his health further, Chifundera gestured for us to move on so that we could leave the airport and commence the difficult journey into the city.

I could not get a clear look at my surroundings in the darkness as we left the Ndjili airport and drove west into the outskirts of the city, where we encountered endless potholes and traffic that crippled our progress. Weary from jetlag and sweating profusely from the miasma of smog and thick humidity, I completely lost track of time as our rented minibus lurched among jarring dips, bumps, and sharp turns in the road. We passed an endless series of roadside shacks and the Congolese equivalent of strip malls. My senses were assaulted by smoke, LED lights, blaring music, shouting, gesticulating, and a noisome mix of car exhaust, food, and sewage. By the time we reached the simple "LosAngeles" Flat Hotel on Avenue Kasavubu in the western part of Kinshasa, I had only enough energy to wolf down a meal of cassava and beef in an oily sauce before dragging myself under a mosquito net to sleep. I would need my rest, because the following day, jetlag or not, we would make a mad dash through the city to buy supplies for our first stop, Bombo-Lumene Reserve, just east of Kinshasa.

Bright and early the next morning I quaffed several cups of coffee to ward off the upside-down feeling of jumping several time zones, and we squeezed into a taxi for a trip into the commercial district of Kinshasa. I greeted the driver and noticed he proudly displayed Congo's blue, red, and yellow flag on the dashboard. As we entered the heart of the capital, Kinshasa struck me as a city of contrasts. Many of the streets we had passed through were full of potholes and eroded pavement, but the main thoroughfare through the center of the city, Boulevard du 30 Juin (in honor of the day of independence in 1960), had four well-paved lanes with clearly demarcated crosswalks and working streetlights. Here and there I could see colonial-era buildings with their distinctive round arches over the front doors, but in between them were modern multistory buildings containing businesses, hotels, restaurants, and government offices. As we passed into the northernmost section of the city, Chif pointed out the five-star Hotel Memling, where journalists had watched Zaïrean soldiers fleeing from the invading forces of Laurent Kabila in 1996.[63]

Everywhere I looked, street vendors sold everything from food to cheap watches under large market umbrellas. African men in brightly colored suits clutched briefcases and cellphones as they walked briskly past less fortunate people pushing carts with various goods they hoped to sell in the city's markets. Large billboards advertised everything from beer to real estate, along with public health warnings about HIV and AIDS. The horizon was full of large cranes that were constructing more buildings, suggesting a buoyant economy.

Chifundera knew the city well, and we left our taxi in an area with lots of shops on the ground floor of buildings that looked like they were built in the mid-twentieth century. Like other cities I had passed through in Congo, Kinshasa was filled with the sounds of blaring rumba music, honking car horns, trucks with broken mufflers (many erupting clouds of black smoke), vendors advertising their wares with megaphones, and people shouting to one another over the din of the city. Men and women greeted one another with handshakes that lingered for minutes into their animated conversations, and I marveled how sleeping babies, swaddled in cloth slings on their mothers' backs, slept obliviously through the racket.

I was briefly introduced to Elvis, our recently hired, eighteen-year-old cook, who shook my hand with a warm smile. He joined Aristote and Wandege on a mission to buy rice, beans, and cooking implements we would need for the first leg of our expedition. Chif and I wandered into several stores to buy malaria medication and other supplies, and later, we ended up at the city's central post office because he needed to retrieve some paperwork from them. We walked past walls with hundreds of postal lockers from the colonial era, but all of them were hanging open and the locks seemed to be missing or broken. I asked Chif if mail service in Congo had been re-established in recent years, because I knew it had not survived long after the collapse of colonialism, and recent packages I had sent to him were mailed to a post office box he rented in Rwanda near the Congolese border.

Chif gave me a sarcastic smile and explained that in theory the Kinshasa post office was supposed to function, but in reality it did not. He told me that recently, he had written a research grant to a funding agency in Japan that required a hard copy in the mail. He took the envelope to the post office, paid the $100 exorbitant postage, and was assured that it would arrive in plenty of time before the deadline. When the Japanese agency failed to confirm receipt, Chif returned to the post office to find out what happened. The clerk at the window had no clue, but behind him on a shelf, Chif spied his envelope, forgotten and rotting in the tropical climate. Somehow he found the sense of humor to laugh.

By the time our errands were finished and we returned to the hotel in the late afternoon, I was running out of energy, but Aristote told me he had captured a lizard during their return from the market, the first one of the expedition. Wandege appeared with a white cloth bag tied in a knot, the "container" of choice for captured lizards and snakes.

"Aristote and Wandege!" I beamed, "you found something *already*?"

Wandege's gentle eyes widened just for a moment in a proud affirmative, and in his measured and heavy accent replied, "Yeeeus, Aristote caught one agamas on zeh wall for you."

Completely forgetting my exhaustion, I eagerly accepted the bag, untied the knot, and peered inside. My glance was met by a foot-long, bluish-gray lizard with an orange head—an *Agama*! Common around the walls and buildings of human settlements across Africa, agama lizards are easily recognizable by their bright colors, at least in males when they are basking in the hot sun, scales as rough as sandpaper, and prominent claws to grip surfaces ranging from brick walls to trees. Studies of the DNA from Aristote's capture would later confirm that this particular lizard was a West African rainbow lizard (*Agama picticauda*).[64] This species has a huge geographic distribution from the western coast of Africa in Guinea-Bissau to Congo, and because of its popularity in the pet trade, it was accidentally introduced to southern Florida, renowned as the state with the largest number of established, nonindigenous amphibian and reptile species in the United States.[65] More than half a century ago, the West African rainbow lizard was identified as the first known vertebrate with temperature-dependent sex determination (TSD), a phenomenon whereby incubation temperature, and not sex chromosomes (i.e., genetic sex determination like in humans), determines the male-female offspring ratio.[66]

As I continued to explore Kinshasa with Aristote on my third day there, it was impossible to miss the contrast of the flamboyant rainbow lizards against the sun-bleached and sooty surroundings. Warmth from the morning sun drove them into a frenzy of bobbing heads, fighting, mating, and scurrying over walls, houses, and huts, contributing to the frenetic vibrancy of the city. Given the infinite number of insect prey attracted to the discarded scraps of food and other human waste littering the streets, I estimated there would be untold thousands of the lizards in Kinshasa.

We finished our logistics shopping spree and decided to escape the punishing afternoon sun by eating spaghetti bolognese and grilled fish at a small café, and knowing that we would not have access to such treats much longer, I drank a couple of ice-cold cokes while Aristote downed two large bottles of Primus beer. Relaxed, happy, and enjoying the scenery surrounding us, we wandered down a small street in the direction of our hotel as I joked with Aristote and asked him which girlfriend he would visit in the evening. He was an incorrigible flirt, and I knew he would seek out female company the moment his duties for the day were over.

Our walking conversation was interrupted by several loud popping sounds that emanated from somewhere down the street. Suddenly, a group of wide-eyed people ran toward us with their hands raised in caution, and shouting warnings in Lingala.

Aristote stopped in his tracks and said, "Oh! Someone was just killed ahead of us. We must go back *now*!"

I felt a shot of adrenaline numb my body, followed by a shockwave of fear. I staggered back clumsily before turning around to retreat. Several people continued to run past us in terror, but Aristote just walked briskly.

"Aristote, do we need to run away?" I asked breathlessly while matching his pace.

"No," he said coolly, "I don't hear any more gunshots, and the killer is probably running away in the other direction by now."

I trusted his judgment, because he had been a commander in the Mai-Mai rebel militia during the war, and he had proved his unrivaled street smarts during our countless previous perils together. I allowed my heart rate to slow down as we walked on to a busy street where the only people running were dashing across the street to avoid traffic. I stopped to take a deep breath, relieved that we were now safe.

"Damn," I said with a glance to Aristote.

"Yeeesssss," he said, in the way he always did when he wanted to emphasize something. "Kinshasa can be dangerous, and so *muzungus* (white men) like you must be careful. I don't need you to die!"

Unnerved, I was eager to leave Kinshasa. My legs tingled from the waning adrenaline as we continued walking to the hotel, now in silence. I could not afford to let the beguiling beauty of Congo deceive me into thinking I was a tourist on vacation. My mind wandered as I contemplated the risks and precautions I was taking, both for myself and the entire team. To ensure I did not have too much cash, I paid my team's salary (half before, half upon completion) via Western Union. To avoid robbery, I took the usual precautions of wearing shabby field-work clothes without jewelry or any other visible valuables, and I never shared my routine or travel plans with strangers who seemed to ask innocently while making conversation. Whenever I had to take out my wallet to pay someone, I tried to do so in a place that was not visible to strangers around me. It probably helped that I was street-smart, weighed over two hundred pounds (much larger than the average Congolese man), and carried around a three-foot metal snake stick (in remote areas) that can be an effective defensive weapon in a pinch. For these reasons and dumb luck, I have never been robbed during any of my eleven expeditions to Congo. But the shooting was a stark cue that all these precautions would be moot if I found myself in the wrong place at the wrong time. And at some point, everybody's luck runs out.

When we reached the hotel, I was pleased to find Chifundera, Wandege, and Elvis waiting for us with the rental truck, and it did not take long for us to cram inside and hit the road to the east. Located about fifty miles east of Kinshasa, we did not have a long journey ahead of us to reach the reserve, but it was already mid-afternoon, and like any road in Congo, there could be unforeseen delays due to everything from accidents to trains of passing cattle herders.

My concerns were justified. We had to retrace our route through the eastern side of Kinshasa along the same terrible and time-consuming road to the airport, but turning north, the traffic eased as the urban landscape transitioned into grassy hills and picturesque farmland and pastures. We crossed a bridge over the Nsele River, one of the many tributaries of the Congo, and stopped to buy some bananas and nuts from a street vendor as we turned east. I grew more excited as we drove over a bridge spanning the Bomba River, where I could see my first glimpse of gallery forest—narrow belts of forest that can only grow along the banks of rivers in drier environments that are dominated by woodlands and grasslands.[67] I knew we were getting close to the reserve, because as the Bomba zigzags south, it eventually splits into the Bombo and Lumene rivers that pass through the western and eastern sides of the reserve, respectively. I knew there would be a large swath of forest sandwiched between the rivers, harboring a flora and fauna that would surely be distinct from the surrounding grasslands.

Dusk found us turning south along the rough road leading to the reserve headquarters, and our bodies jostled as the truck negotiated the rugged terrain. We encountered a small group of rangers dressed in green uniforms who waved us through a gate. When the driver lowered his window long enough to exchange a quick greeting, I could hear a chorus of dozens of chirping crickets seeking mates in the vast savanna around us. At the end of the road, we unloaded all our gear into a grassy clearing while a blizzard of flying insect bodies swarmed around our headlamps and probed every orifice they could find. Congolese grasslands are especially well endowed with insect life, and to avoid the hordes, we hastened to move everything into a set of whitewashed buildings that the rangers offered to us as our living quarters. I noticed that our hosts were not using any flashlights, perhaps because good batteries are nearly impossible to find in most of Congo, but then I wondered how they avoided stepping on venomous snakes in the dim light.

Covered in dust and sweat from the drive and luggage transfer, I took a long swig of lukewarm water as I assessed my living quarters, which were typical for small towns and outposts in Congo. On the plus side, the building was about fifteen by twenty-five feet, more than enough space for me and my equipment, and the security was ensured by bars on the windows and a sturdy metal door with a lock and key. Although some of the glass louver windows were cracked, they seemed to be intact enough to exclude the legions of insects I had encountered outside. However, I could see plenty of spider webs in inaccessible corners near the ceiling, many laden with the carcasses of their victims. A few moths fluttered toward the bright beam of my headlamp, and I noticed a bustling metropolis of ants, beetles, and other bugs on the floor. There was no electricity, the concrete floor was dusty, and the bathroom had a seatless toilet and mildewed bathtub, but no running water. To compensate for the latter shortcoming, a sizable bucket of water had been left next to the toilet, a nearly universal necessity outside large cities in Congo. One could always judge the age of such courteous offerings by the number of drowned flies at the surface, or in the case of more stale water, the development of mosquito larvae. A couple of well-used twin beds were available for guests, but based on years of experience in the Congo, and to ensure adequate protection from malaria-carrying mosquitoes and other biting insects, I deployed my one-man tent, air mattress, and sleeping bag on the floor. The honeymoon of hotel rooms with air conditioning and comfortable beds strewn with mosquito nets was over.

On the seventh of June I awoke at dawn to a hazy sky so thick that it made the sun look like a dim orange headlight straining to illuminate its path in fog. In front of me a grassy clearing contained several palm trees and the African version of a pergola—tall logs of wood supporting a rectangular thatched roof in a hut-like umbrella, providing shade to a table surrounded by plastic lawn chairs. Such structures are quintessential hallmarks of African villages that are used for meetings, meals, and other social gatherings. Aristote told me that in Congo, everyone calls them *perruque*, the French word for wig. Just a few feet beyond it at the edge of the clearing, I noticed the distinctive reddish leaves and thick bark of the peeling-twig combretum tree, and a sea of light green grass and waist-high shrubs between them. As I moved closer to the trees, I realized that I was standing at the top of a huge hill on the southern edge of the Batéké Plateau that looked down into the Lumene River valley, perhaps two hundred feet lower, where I could see thick ribbons of rainforest trees hugging the path of the water. In the distance, several more rolling hills of greenish-gold grass continued for miles into the horizon.

Bombo-Lumene Reserve was originally created as a game reserve in 1968, mostly for expat sport hunters from Kinshasa, but due to its ecological and conservation importance, a large portion of it was elevated to a more important International Union for the Conservation of Nature (IUCN) reserve (allowing some sustainable use of its natural resources) in 1976.[68] I imagined an army of animals hiding in the brush and wondered whether any large mammals still survived in the reserve. At the height of the Kongo kingdom, perhaps in the very spot where I was now standing, houses were built with palisades and bolstered with walls of thorny palm branches to protect them from lions. I knew that African lion (*Panthera leo*) populations had nosedived throughout most of their historical geographic range, disappearing from much of Central and West Africa in the last few decades,[69] but could a small pride still lurk in the grass below me? Did I need to worry about being *eaten*?

Aristote coughed behind me to ensure he would not startle me and then asked "How did you pass the night" in his rough translation of the Congolese French expression *Comment s'est passée la nuit*?

We exchanged morning pleasantries, and then I gestured to the scenic landscape below us. "Think we will find something good down there?" I asked.

"Yeeessss," Aristote replied. "I think there must be something good there, and you know we are ready to find many animals!"

It was then that I noticed he had two clear plastic containers in his hands, which we utilized for temporary housing of our smaller captures. Aristote observed my glance, offered the first container to me, and said, "These geckos were found in huts in the night."

I could see three pale lizards in the container, and given their locality in a human dwelling and their ghostly color, I knew they were African house geckos (*Hemidactylus mabouia*). Ubiquitous in any human habitation in Congo, as well as much of sub-Saharan Africa, these nocturnal, finger-length lizards can use the special adhesive lamellae on their toes to scale walls in search of insect prey. Although most of the circa two hundred species in the genus *Hemidactylus* have relatively small geographic ranges, at least ten species have managed to spread to other continents, including the African house gecko, which likely hitchhiked in slave ships to establish populations in the Neotropics by the mid-seventeenth century.[70] If you live in the southern half of the continental United States, in many places from California to Florida, you have probably noticed the invasive, salt-and-pepper patterned Mediterranean house gecko (*Hemidactylus turcicus*) hunting for insects on walls at night, and models suggest they are likely to spread north in the future as climate change warms the planet.[71]

When my eyes moved to the second container in Aristote's hand, he said, "Chifundera found these three lizards in his bathtub last night."

A quick glance at the lizards' shiny scales and long tails suggested they were skinks, one of the most diverse and globally widespread groups of lizards on the planet. Upon closer inspection, I noticed their distinctive brown bodies with a heavy sprinkling of pale white flecks, orangish-red flanks, and canary yellow throats and lips. I identified them as the speckle-lipped mabuya (*Trachylepis maculilabris*), one of the most common lizards in nonforested areas of Central Africa, especially near human habitations, where they chase after their insect prey. Preserved muscle tissue from these individuals would be used in a 2019 study of *Trachylepis* skinks to show that some populations of the speckle-lipped mabuya are genetically distinct from others, a pattern seen in many other widespread species of amphibians and reptiles in Africa, including African house geckos.[72] Morphological study (size, shape, color pattern, scale patterns) of natural history museum specimens from these populations is essential to determine whether the genetically distinct populations are merely minor variants of one species, or so-called cryptic species, which look similar, but have enough distinctive differences (e.g., DNA, morphology, ecology, behavior) to warrant recognizing them as distinct species.

It can be simultaneously exciting and frustrating for taxonomists, including me, when populations like this are in the gray area between valid species and minor intraspecific variation. Scientists must provide as much evidence as possible to convince the global scientific community of their taxonomic decisions, a labor of love that often demands meticulous examination of many museum specimens, their associated DNA samples, and a vigorous investigation of all previous research on the animals. Most of my career has been dedicated to this detective work, because it is an essential prerequisite to understand each species' role in the dynamic complexity of the biosphere, Earth's global ecosystem of living organisms and nonliving components and processes (e.g., climate, water cycles) that make life possible. The most rewarding aspect of this endeavor occurs when the accumulated evidence points to identification of new species to science, and in the intuitive human process of naming them, scientists make the world aware of these unique evolutionary lineages, which deserve recognition and inclusion in conservation efforts. This exciting potential for new species discoveries has drawn me to Congo since 2007, where basic herpetological knowledge has an especially large blind spot in need of more study.[73]

Pleased with the first few lizards, but knowing that more exciting discoveries awaited us, we ate Elvis's scrumptious breakfast of fried plantains and rice, and descended the steep path into the valley beneath us with a couple of armed rangers for protection. I peppered them with questions about large mammals and learned that many lions once inhabited the area, but in 1973, Mobutu responded to complaints from local people (probably losing livestock) by sending his soldiers to shoot them all. Because of overhunting in the reserve, the numbers and diversity of herbivorous mammals declined over the years, reducing the prey base for large carnivores, and they did not think leopards were present either. The habitat seemed perfect for savanna elephants, but the rangers claimed there had never been any there. This was hard to believe, but given the proximity of the reserve to Kinshasa, it was likely that Belgian hunters had wiped them out many decades ago, long before the collective memory of the rangers. Nonetheless, I was glad to have the rangers' protection, because they told us hyenas were still present, along with potentially dangerous forest buffalos. The reserve still had plenty of medium to small-sized mammals, including several kinds of antelope, hogs, civets, genets, mongooses, and monkeys.[74] We could see and hear several kinds of birds, including a large flock of helmeted guineafowl (*Numida meleagris*), a bit larger than the average chicken and easily recognizable from their grayish-black feathers with white spots, pale blue head, red cheeks, and light brown crests that resembled beanie hats.

As we approached the Lumene River, I could see impressively tall, old-growth trees on the far side of the river's dark brown water. The rangers guided us to a bridge made of intricately woven strands of lianas, vines, small trees, and rattan fiber, which spanned the river. The structure looked like a monument to a previous century, in the exact style of the Bakongo, who constructed "royal routes" for the passage of kings since time immemorial.[75]

The bridge swayed under our weight as we passed over it, necessitating the need to steady ourselves by grasping the meshwork of vegetation on the sides of the structure. But ants were also taking advantage of the pathway across the river, and their bites encouraged us to hasten our crossing to the other side. There we found ourselves in dense semi-deciduous forest,[76] distinctively dimmer than the surrounding open grassland, but not so thick as typical rainforest that we would encounter farther north along the Congo's tributaries. Relatively cool and humid, it was easy to understand how species adapted to the rainforest might live here, but not in the dry and hot grasslands. After hours of searching, we found only a handful of small green reed frogs (*Hyperolius*) that were sleeping under leaves in a marsh. We enjoyed the seemingly endless parade of bright orange butterflies, and one of the rangers brought our attention to some buffalo footprints in the mud.

Once dusk fell and the night crew of animals became active, our luck changed. In the forest we found strange-looking toads (*Sclerophrys*), fist-sized white-lipped frogs (*Hylarana*), rocket frogs (*Ptychadena*), and a tiny squeaker frog (*Arthroleptis*). When we passed through the gallery forest on our way back to the grasslands, we searched a stream and found some aquatic frogs that proved to be a new species, which several colleagues and I named the Gabonese clawed frog (*Xenopus mellotropicalis*) after we realized it had a large geographic distribution, including Gabon.[77] When we returned to the marsh, we found a boisterous chorus of several frogs calling for mates, including more green reed frogs, puddle frogs (*Phrynobatrachus*), kassinas (*Kassina*), and many others. I braved clouds of insects to record the mating call of some of the reed frogs, a team effort requiring everyone to stand still in silence and darkness while I stretched my arm with a microphone toward the vocalizing individual. After completing each recording, I turned on my headlamp long enough to record the time and temperature, look for more calling males, and admire the purple flowers of the hooded harlequin orchids that bloomed along the edge of the marsh.

But the most interesting find of the night was a gecko that we encountered in grass near the river. It was a *Hemidactylus*, but unlike the African house gecko, it had a striking pattern of squiggly dark brown markings and chevrons, gray spots, and a series of white spines on its flanks. A second individual was found in grass near the reed frog pond. Years later it would be named as a new species, *Hemidactylus gramineus*, the species epithet referring to its apparent affinity for grass, a highly unusual habitat for such a gecko.[78]

Over the next couple of days, the rangers continued to help us find several additional species, Elvis provided high-quality meals, and my camera and laboratory were kept very busy documenting every detail of the discoveries. Just after dark on my final night in the reserve, a strange light caught my eye in the distance, and I saw a wall of fire consuming the grasslands on one of the hills of the horizon. Although fire caused by lightning strikes is an essential factor in maintaining grasslands naturally (otherwise woody plants take over), I suspected somebody was purposefully burning away the vegetation to promote new growth for livestock. Because of multiple threats caused by humans, tropical grasslands are one of the most threatened terrestrial biomes on Earth—a staggering 70 percent of the planet's grasslands and savannas have been transformed to croplands or grazed landscapes.[79] The fire was a stark reminder that at least some of the species I was encountering would probably not survive the twenty-first century. With a ballooning African population that is on track to double in size by mid-century to 2.5 billion people,[80] I contemplated how long the reserve's rangers could hold back the tidal wave of human pressures.

I continued to stare at the fire, mesmerized by its light, sinuous movements, and unpredictability. My mind wandered to the dangerous itinerary that I had planned for the team. In the morning we would return to Kinshasa, buy more supplies, and move closer to our ambitious river journey, which would take us hundreds of miles along the Congo River, its tributaries, and beyond them into no man's land. Even if we returned from the journey without a scratch, how would our experience change us?

I lulled myself to sleep by doing what I always did in Congo when I got worried. I accepted that I was hurling us into the unknown, but I could not predict or control the misadventures ahead, and thus, it made no sense to waste time worrying unless disaster actually struck. If only this superpower worked outside Africa.

RUNNING MALUKU'S GAUNTLET

I could smell the Congo before I could see it. Dank and resembling moldy earth after rain, the mugginess of the river mixed with its odors in a cocktail that seemed to coat the inside of my nose with a lingering film. Thousands of miles of river had carried the detritus of countless plants, animals, fungi, bacteria, and humans to Malebo Pool, where it slowed and mixed with the pollution of Kinshasa in an unmistakable brew. It was too dark to get a glimpse of the river when we arrived at the Auberge Oli-Palace "hotel" in Maluku, a commune of Kinshasa's northern corner, the night before our river journey would begin. The distinctive scent and palpable increase in humidity forced me to internalize the imminent trip upriver. I was mostly excited, but also a little anxious.

The "hotel" was actually an L-shaped cinderblock building with several tiny rooms that included twin beds with mosquito nets, which had seen so many punctures, rips, and tears that they looked like rags of swiss cheese. I struggled to cram my baggage into the prison-cell-sized room and then set up my one-man tent on the bed with my headlamp. Heat trapped by the corrugated metal roof made the room uncomfortably warm, but it would be unsafe to prop open the door (the only opening into the room) after dark in this part of the city. Sweat dripped off my body as my mind raced with thoughts of the voyage ahead, and I slept very little.

I emerged from the room at dawn and noticed a couple of stray dogs that were sniffing around for scraps and a middle-aged woman in the courtyard who paused from her stooped sweeping with a long palm frond to stare at me. She was dressed in a dusty green *kitenge* (the African equivalent of a sarong) with a matching green head wrap, the colorful attire of middle-class women in Central Africa. Deep wrinkles etched her face and her mouth hung open slightly, as if my presence had frozen her in place. I could understand her surprise—the Auberge Oli-Palace probably would not have rated well on Tripadvisor if it had been listed there, and I guessed from her reaction that I was probably the hotel's first white guest.

I waved, smiled, and in my best Lingala greeted her by saying, "*Mbote*, mama (Hello, mother)."

Her body jerked up and she returned the wave with a wide smile, revealing several gaps of missing teeth. I retired to my room to pack up the tent and wait for my companions to wake up.

Eventually, Elvis showed up with some hot water for my instant coffee and a modest breakfast. Chifundera dropped by to tell me that he would contact the captain of our boat to get transportation to the bank of the river and I should wait "some minutes" at the hotel to avoid thieves. Familiar with the languid pace of life in Africa, I opened a book and read while I waited.

Two hours later, Chifundera showed up with the boat crew in a minivan (a.k.a. bush taxi) that looked like it had been through a war. I hurried to settle the 30,000 Congolese francs ($33) tab for our five rooms. We jammed our baggage into every orifice of the van until it was full, but when I tried to open the passenger-side door next to the driver, the only place where my large frame could fit, he gestured in French, "Monsieur, ça ne marche pas (Sir, it does not work)." Unfazed, I entered "Dukes of Hazard" style by wiggling my body through the open window. With all his passengers accounted for, the driver gave a command to two of the crew, who commenced pushing the van. Puzzled temporarily, I watched as the van rolled forward toward the top of a steep hill leading down to the river, but the engine was off.

Once we started rolling down the hill from gravity, the driver kickstarted the van with a sharp grating noise, the two crew members jumped in, and we started our descent toward the Congo. As our momentum increased, I noticed the brakes squealed and the suspension was all but shot, and our vehicle careened over rocks, potholes, and large tree roots with jolting violence. We seemed to be descending a narrow alley in between densely populated neighborhoods of dilapidated shacks and cinderblock buildings with shops, bars, and small restaurants. I caught short snippets of rumba music as we passed the bars, but it was hard to hear anything over the din of our vehicle's collisions. Everyone in our path quickly darted to the side of the alley, sometimes with cursing gesticulations, and other times with bursts of laughter. I glanced back at my fellow passengers and saw their concerned expressions and death grips on the seats. Elvis held his head in his hands and seemed to be praying. Somehow we reached the bottom of the hill without popping a flat tire, and I could finally see the bank of the Congo ahead of us.

When our van came to a creaking stop at the edge of the beach, I exhaled a deep sigh of relief, and Aristote exclaimed in Swahili, "*Asante mungu!* (Thank you, God!)." But the driver and crew did not skip a beat as they poured out of the van to carry our baggage down to the boat. I extracted myself through the open window onto muddy soil that was covered in trash. I saw about two dozen wooden boats of various sizes, a large market, and tiny shacks hawking everything from refrigerated plastic bags of water to sweatpants. The Congo's smell had become more acrid here, and it was mixed with fires from street vendors cooking everything from fish to monkeys.

I helped carry some equipment to the bank, where I scampered up a rickety gangplank on to a thirty-foot wooden boat with several other passengers, including Wandege and Aristote. We accepted plastic lawn chairs offered to us by the crew, and I made myself comfortable under the merciful shade of a canopy that covered most of the boat. I noticed an inch or two of water that had pooled in the hull, but I dismissed it as splashes from the river. Soon we would leave Kinshasa and travel north to Kwamouth (pronounced *kwa-moo-too*), a town about one hundred miles (as the crow flies) to the north.

Like countless other places in Congo, Kwamouth has a troubled past. One of the first missionaries to reach the site in 1886 was Reverend Father A. Merlon, who fumbled into a serious misunderstanding with the local tribe when he set a survey marker at the edge of their field. Thinking it was some nefarious spell, the Congolese beat their drums in alarm, and Merlon was quickly surrounded by "démons" who threatened him with arrows, spears, and war knives. In response, he calmly picked up his gun, shot the opposite bank of the mile-wide river, and told the shocked chief and onlookers that the white man's gun could strike them everywhere. Merlon then told them that unless they retreated immediately, he would eat them, their children, subjects, goats, and crops, adding he would still be hungry afterward. He did not experience any trouble after that.[1] Soon thereafter in 1888, the Congregation of the Immaculate Heart of Mary established their first mission at Kwamouth, but within three years, half of them had died of sleeping sickness.[2]

My interest in Kwamouth was only as a temporary stop at the confluence of the Congo and Kwa rivers, en route to more remote places. I looked forward to the interesting scenery we would encounter during our journey, which Chifundera guessed would take about eight hours, depending on the current and power of the boat. I glanced at my watch—it was 9:30.

Chifundera appeared at the bow of the boat and gestured for me to join him on the beach. He had a look of concern on his face. "Eli, please give me the money for Elvis's salary until now," he said.

"Why didn't we take care of this before?" I asked, assuming he wanted a last-minute advance for his family. "He already had a good-faith payment at Bombo-Lumene, and you know I don't like to dig out money in front of large crowds like this."

"Yes, I know, but Elvis has quit," he replied. "I think his father scared him by telling him the journey would be dangerous and maybe he would not come back."

I understood how Elvis could lose his nerve, but his timing could not have been worse. It was telling that he was nowhere in sight to say goodbye. Gobsmacked, I just stared at Chifundera in horror for a moment. He returned my look with wide eyes to acknowledge our predicament.

"But we already have everything packed to leave, what are we going to do?" I asked.

"We must go, of course," he replied with a dismissive wave of his hand. "We will have to find another cook along the way."

In the shock of the moment, I was not thinking clearly, but I knew it was not a good idea to leave on a remote expedition without a cook to feed the team. I hesitated for a moment, but having no choice, I grabbed my wallet and produced the money needed to pay off Elvis. I felt uneasy as my actions were observed by several passersby. Still reeling from disappointment, I turned to retrace my steps to the boat, but Chifundera asked me to wait for him on the beach.

He returned after five minutes and said, "Elvis says thank you and sorry."

"Yeah," I mumbled.

"Do you have your passport?" he asked.

"Yes," I replied. "What now?"

"Come with me. We must see immigration before you can leave."

I rolled my eyes, knowing that this would be a headache. The Direction Générale de Migration (DGM) was notorious for creating problems for foreigners in Congo, especially when crossing borders. But even if one was moving from one town to another within the same province, they had the right to ask for a passport to ensure everything was in order. They excelled at creating problems that could only be solved by long arguments or, as they really hoped, a fine payable in cash to them. One common tactic was to claim the visa to enter the country had expired, even if it had not. Another was to demand proof of a vaccination that nobody had ever heard of. Some of them were genuinely concerned that I was in the country legally, but others were in search of shady ways to make these seemingly serious infractions go away. Over the years, my Congolese colleagues and I had been amazed by their complete lack of compunction and endless stores of energy to argue in favor of their spurious assertions. Now I had to cross my fingers that we would encounter honest men.

Feeling like I had been called to the principal's office, my head hung low as I entered a dim shack with two sweaty DGM agents sitting behind a desk. A photo of President Joseph Kabila hung on the wall beside a calendar, and several notebooks, stamps, and stamp pads were scattered across the ink-stained table. I did my best to answer their questions in my broken French, and Chifundera produced a small pile of permits from his Congolese research institution to justify our trip. When they handed my passport back to me without any major objections, I became hopeful that we had dodged a bullet. But then they told Chifundera that he needed this and that permit for us to leave. As I was about to turn to leave, the head agent looked at me seriously and said "Bon voyage" with a sarcastic smirk.

After we exited the shack, Chifundera reassured me that all the offices issuing the required permits were sprinkled around the beach, and he would try to get them quickly so that we could leave. "I think it is best for you to go back to the boat and wait with Wandege and Aristote," he suggested.

I was just about to comply when our heads were suddenly turned by the shouts of women just fifty feet away from us. Several of them were screaming in anger at a young woman, who was shouting back at them with defensive gestures. The police showed up and arrested the target of the women's ire, and she burst into tears as they dragged her away. As she was escorted through the crowd, the police had to beat back some people who tried to assault her. The young woman howled in anguish and her body went limp in the hands of the police.

Chifundera broke the stunned silence between us by telling me, "The people are saying she killed her baby." Too disturbed to reply, I stared at the angry mob and wondered what kind of justice the woman would face. It was an unsettling sight and a bad omen for the forthcoming trip.

I returned to the boat to find a small group of officials yelling at the captain and crew, because they wanted additional money for various fees, taxes, or documents. Apparently, word had spread that they had a white passenger, and everybody with a smidgeon of authority was taking advantage of it. One of the annoyed crew members shooed me on to the gangplank, and when I returned to my lawn chair, I told Wandege and Aristote everything that had happened on the beach. They had been talking with some of the other passengers, who informed them that the eight-hour timeframe of the journey to Kwamouth was a gross underestimate. I sank back into my chair and wondered if the bad news would ever stop.

A few more passengers and cargo were loaded on board as two hours passed by with no apparent progress toward our departure. The arguing between the captain and officials on the beach was incessant. Given the bad news about the time required for the journey, I was beginning to get concerned that our late departure would mean arriving in Kwamouth after dark, making the trip more dangerous. Moreover, I was sure the mosquitoes would be horrendous in the marshy environment of the Congo after dusk.

When Chif returned to the boat, I shared the bad news about the longer trip time, but he was not surprised.

"Yes! The boat ride will take one day and a half," he said nonchalantly, as if this had been the original plan all along.

"WHAT!?" I cried. A journey that I thought would take all day would actually be a miserable full day and a half, including a night on the river. I was completely unprepared for that prospect, and for a moment, I started to panic.

"Are you sure we can't find a faster way to Kwamouth?" I asked.

"Yes, I am sure!" he said, growing impatient with me. "You have already paid our fare and after all the delays from us, there is no way they can refund the money if we cancel. And if we miss this boat, it can take several days to find another one." I could not argue with him, because even with ten weeks for the expedition, I wanted to spend as much of it as possible in the forest.

At 12:30, we finally pushed back into the Congo, executed two slow 360-degree turns, and prepared to turn north. But there was even more emphatic shouting from the bank, and we immediately returned. After additional arguments and shouting, more passengers and cargo flooded on to the boat until there was very little room for anyone, even with some people perched on the roof of the canopy. Another excruciating hour slipped by until we left again. We had traveled only two minutes away from the port before a speedboat flying the blue, red, and yellow star flag of Congo pulled up to the side. I felt a surge of fear as a skinny soldier with an AK-47 jumped on board, shouted in fury to the crew, took one of our jerry cans of fuel, and sped off to the bank. It seemed that somebody had not been paid, and they were taking the fuel to prevent us from leaving.

We returned to the bank yet again, and Chifundera and the captain hopped off the boat to sort out the latest problem. I waited twenty minutes, but feeling antsy about the dismal circumstances, I decided to look for Chif to tell him I had enough and we should leave.

Aristote and I caught up with him outside one of the shacks that served some kind of official purpose. He was engaged in a three-way argument with the captain, an official behind a desk in the shack, and a small group of soldiers. Clearly agitated, he gestured emphatically as he clutched a fistful of paperwork. I signaled for him to join me on the sidelines for a moment.

"Chif, I don't know if I can do this. Maybe we should look for another way to Kwamouth," I said in a hangdog voice.

Now it was Chif's turn to be shocked. In an eruption of exasperation, he said, "Look! Just look at these documents I have been getting for you!" With a wild swing of his arm, he showed me official documents giving us and the captain permission to leave, and in an annoyed voice said, "Look, count them! ONE, TWO, THREE, FOUR, FIVE, and now I am trying to get the SIXTH! The captain must get these signed, stamped, and paid for so that we can go. If we leave now, it will be the same thing tomorrow, but you will have to start from the beginning!" His expression had an intensity that was impossible to contest.

As I looked at the pile of paperwork in his hand, I knew he had been working hard for hours to get them. I had to concede he was right, but I could not shake the feeling of being trapped. I hesitated only a moment before I replied, "Ok, do it." My head hung low as I walked back to the boat, now completely resigned to a very long and uncomfortable trip north. Along the way, my grumbling belly reminded me that I had eaten only a small breakfast many hours before, and given the revelations about the long trip ahead, we would need more portable food. Aristote and I bought several bags of grilled goat meat and chapati (a greasy flatbread) for the team's lunch, dinner, and breakfast.

It was after three o'clock and sweltering when we left Maluku for the last time, but just as we were about to pass the last docked boat, I saw a man in a full-length woman's winter coat clinging to its bow. He gestured to the captain, who angled our boat toward him, allowing him to hop into our bow as we passed.

He landed with catlike agility, and as he steadied his feet and stood up, everybody stared at him because his complexion was noticeably darker than everyone else's. I wondered how he could stand the heat while donning a heavy coat.

Wandege, Aristote, and I sat on our lawn chairs in the stern, where we were able to climb over a minimum number of people to answer nature's call off the back of the boat during the river passage. Chifundera opted to socialize in the bow. I sat across from a woman in her mid-twenties with two young boys, one about three years old and the other an infant. The eldest boy stared at me curiously, and I smiled at him and his mother.

Leaving Kinshasa behind, I started to see a series of abandoned and rusting boats on the bank, some of them enormous and modern, whereas others seemed to be a century old. I looked down at the pool of water near my boots and noticed that its depth had doubled since we left Maluku. Following my gaze, one of the crew noticed the worsening leak, too, and seeming to expect this, he stepped forward with a bucket to fix the problem. Once every hour from that moment until the end of the journey, one of the barefoot crew would bail out the water that accumulated in a large depression right next to my chair, occasionally splashing me as the bucket tipped overboard.

As we continued north, we passed an endless series of small fishing villages, each with a few dozen huts and dugout canoes docked at the bank. Every hour or so, we stopped at one of them to exchange passengers, deliver goods, or give everyone a chance to buy food or water. No wonder it would take a day and a half to reach Kwamouth, I thought.

I noticed Chifundera laughing with the man in the winter coat, and after an hour of conversation, they carefully maneuvered around the throng of passengers to join us in the stern.

With a delighted smile on his face, Chifundera introduced Poignard to me. Now that I could see him up close, he struck me as a dead ringer for a forty-year-old Wesley Snipes. Shorter at about five feet, six inches, his body was built like a running back football player, and he seemed to be exceptionally fit for his age. Unlike the closely shaved heads of every other Congolese man on the boat, Poignard coiffed a box cut. But the truly exceptional character of his appearance was his eyes. Dark and penetrating, his gaze was unusually intense, giving the impression that nothing could escape his notice.

He extended his hand to me with a smile, saying in a gruff voice, "Bonjour, Monsieur Eli."

"Ravi de vous recontrer (Pleased to meet you)," I replied.

Chifundera explained that Poignard had greeted him in Mashi, somehow guessing that they were from the same Bashi tribe in eastern Congo. When Chifundera explained what we were doing, and how we had just lost Elvis, Poignard volunteered his services as a cook.

Slightly suspicious but having no alternative, I agreed to hire Poignard as the expedition cook on a probationary basis. After we reached Kwamouth, we would try out his food, and if acceptable, he could continue with us until the end. We negotiated a salary, and everyone seemed to be very pleased and relieved. Perhaps at last our luck was changing.

I watched the scenery of the grassy hills and abundant bird life until it grew dark, and then I doused myself with mosquito repellant and put on a head net (to block biting insects) that amused several of my fellow passengers, especially the three-year-old. Because we were behind schedule, our boat chugged on for hours in the silvery glow of a crescent moon. We finally docked on the bank of one of the small fishing villages at ten, and I repeatedly failed to contort myself into a semi-relaxed sleeping position on the lawn chair. As the miserable night wore on, the mosquitoes were relentless, and the temperature dropped until it was surprisingly cold. As I drifted in and out of sleep, I was too tired to notice when it started to rain, but a stampede of people escaping it caused us to be crammed together more tightly. At least with so many people next to me, I judged optimistically, it would be nearly impossible for a thief to pick my pocket in my sleep.

At dawn, when the stirring passengers around me ended my fruitless attempt to sleep, I was glad to end my fitful night trying to stay warm. Imagining dragons in the mist around me, suffering a caffeine-starved headache, and catching a stare from Poignard, I was exhausted and eager to move on to Kwamouth. I heard a strange tinkling sound and turned my head to see the three-year-old urinating into the pool of water near my boots. He seemed to be very proud of his aim, and his innocent smile made me laugh. My companions started laughing, too, and Aristote joked that he was now a man.

With a few shouts to rouse any passengers who had spent the night at the riverbank, the captain fired up the Yamaha Enduro engine at five, and we continued our route north. The chapatis had made a satisfactory breakfast, but the greasy carbs did not sustain us for long, and by noon, we were hungry and out of food. For hours, we passed countless fishing villages, and desolate hills with a mix of scrubby vegetation, squat palm trees, and tan earth. Occasionally, we passed small patches of rainforest trees that reached the bank, hinting we were entering the forest-savanna transition zone.

Wandege became so intrigued with a fisherman who was using an orange tarp as a sail on his dugout that he leaned over the edge of our boat to take a photo. As the hours dragged on, I fantasized about coffee, food, and a bed. Bored, I took out my global positioning system (GPS), a phone-sized electronic device, to see our position on a map and track our progress. I was dismayed to find that we were moving along at only three to four miles per hour, and even this was an overestimate. Of course, we stopped at every small village to exchange a passenger, buy food, or get in a random argument for forty-five minutes at a time.

About twelve hours later as we passed through a narrow stretch of the Congo that was only half a mile wide, we were just six miles away from our goal at Kwamouth when the engine began to sputter. Apparently one of the teenage crew had neglected to buy more fuel at the last village, most likely because he was engaged in an argument, and now we were dead in the water. The captain called him a chimpanzee at the top of his lungs, but another member of the crew had squirreled away a whiskey bottle with a little extra fuel. Everybody on the boat was silent with anticipation as we desperately searched each bend in the river for the next village and the possibility of fuel, but it was not long before the boat's motor stalled again. The sudden silence of our engine, coupled with the gentle waves of the river, marked our rapid transition from hope to despair.

I could not stand the thought that we would spend another miserable night on the Congo, and if another boat did not come along to rescue us, we would probably be stranded for several hours the following day. The crew used oars to lead the boat into a narrow channel that was surrounded on both sides by a wall of reeds and water hyacinth, and as we passed a small dugout canoe with a fisherman and his son tending a net just next to us, a vigilant man on the canopy yelled the magic word—*nyoka* (snake)! Following a shriek of horror from an old woman, everybody stood up to peer over the starboard side of the boat to look for the source of the alarm.

It was indeed a snake, but it was not moving, and I guessed it had drowned in the man's net. Chifundera stopped the boat long enough to buy the animal for 2,000 Congolese francs, a little over $2. If we had not come along, the fisherman would have likely cooked and eaten the snake, because they are a good source of protein. My dejected spirit rose, because a snake encounter—any snake encounter—is always a cause for celebration in the herpetology world, especially in the tropics where their diversity is highest. One would think that snakes occur in great numbers in the African tropics, and in some areas, certain species are indeed relatively common.[3] But in the heart of Congo, very few of the dozens of known species are common, and one is just as likely to encounter something extremely rare. Some snakes from this poorly known region, including Laurent's file snake (*Mehelya laurenti*), are so rare that they have only been found once.[4]

Chifundera offered a dilapidated oar to the fisherman, who gently draped the serpent's body over the end of it. After maneuvering it on to the boat and past half a dozen people who grimaced as they leaned out of the way, he deposited it gently on someone's baggage near me. Although the snake was clearly dead, I had to be very careful handling it, because if it was venomous, its fangs might still contain enough deadly venom to kill a man if one were careless enough to bump a finger into them. I did not immediately recognize the species, but it looked suspiciously similar to the banded pattern of a water cobra, a highly venomous denizen of Central Africa's lakes and rivers. About a foot and a half long, the snake had a brown back that transitioned into cream flanks, and overlaying this ground color, a series of dark brown bands passed horizontally over the animal's back, where each one bifurcated on its downward path toward the belly. The brown head had a snout that seemed too tapering to be that of a cobra's, and when I pried open the mouth to look at the teeth, I did not see the telltale fangs of a venomous snake.

I concluded it was a harmless ornate African water snake (*Grayia ornata*), one of only five species in the genus.[5] Recent phylogenetic studies (akin to family trees derived from genetic data) have been inconsistent about the closest relatives of African water snakes, but surprisingly, the best estimates suggest they are cousins of the Calamariinae subfamily, a group of Asian forest snakes.[6] *Grayia* sometimes live in the same freshwater habitats as highly venomous water cobras in Central Africa. The color patterns of both snakes are superficially similar, and both of them have a tendency to end up in fishermens' nets, so it was wise to be cautious with my examination. Reaching a maximum size of five feet, the ornate African water snake mainly feeds on fish, and they lay eggs on land to reproduce.[7] This species also has cultural relevance to some African tribes. According to the folklore of some ethnic groups in Gabon and Republic of the Congo, parts of the snake are believed to improve swimming and fishing abilities, treat sprains, splinters and problems during childbirth, and expunge sickness from newborn babies. The Nzebi of Gabon also believe that if a person is bitten by this snake, a rare occurrence from this docile species, they will be protected from all snakebites in the future.[8]

I was delighted to have the first snake of the expedition, but my elation quickly wore off as the reality of our predicament sunk in. For hours, we utilized a combination of rowing and short spurts from the dying engine to approach a small village that was hidden away among a maze of reeds and islets. Thinking we would spend the night there, at least off the boat, I waited impatiently for the staccato conversation in Lingala between the crew and villagers to subside. Finally, an hour later as the Congo's chilly mist returned, the crew decided we needed to change boats before continuing on to Kwamouth immediately!

It took another thirty minutes to transfer all the remaining passengers and cargo to a dugout canoe with an outboard motor, and as we waited to leave, it listed precariously. A sliver of moonlight penetrated the clouds long enough for me to see that we were surrounded by imposing walls of vegetation, and tantalizing "tink-tink" calls of *Hyperolius* reed frogs seemed to emanate from every nook and cranny of it. Before, the size and girth of the larger boat had been a barrier to most of the elements, but now, the edge of our dugout was only a few inches above the level of the water. My mind drifted to thoughts of crocodiles, and I pulled my arms away from the edge of the boat.

Chifundera wisely produced the life vests we had not bothered to unpack before, and they had the pleasant benefit of keeping us warm as the temperature plummeted. There we waited, limiting our movements to avoid capsizing, until an hour and a half later, the captain returned to the boat with fuel. The motor fired up, and we retraced our path from the backwater in the reeds to the main channel of the Congo. I glimpsed the lights of a small town called Ngabe on the west side of the river in the Republic of the Congo (a.k.a. Congo-Brazzaville), and contemplated how Congo's neighbor to the west had the benefit of power for its towns, whereas most of the places on our side of the river were in complete darkness.[9]

Our pace seemed to be faster as we cruised north into a brisk headwind that buffeted our bodies, causing us to bow our heads and shiver. As we drew closer to our destination, ghostly figures of lone fishermen blinked in and out of sight as they adjusted their flashlights while standing on tiny dugout canoes. Nocturnal fishing must be especially rewarding, I thought, because an accident in the middle of the Congo at night surely meant a quick death.

At last, bright lights could be seen ahead, and we disembarked on to the beach at Kwamouth just after 10:30. I could hear the droning of a gas-powered generator nearby, but everything else was silent. We still had to figure out where we would stay, no small feat at night, but we had finally reached dry land, and our ordeal was now over. Our one-hundred-mile boat trip from hell had taken thirty hours.

As I waited with the baggage for Chifundera to inquire about a hotel, I contemplated how the Congolese people confront adversity with boundless reserves of patience and fortitude. Our boat experience, unfortunately, was typical of the difficulties attached to any undertaking in Congo. Nothing works the way one expects it should in other parts of the world, everything takes an order of magnitude longer to accomplish, and red tape combined with a heavy dose of corruption creates the perfect recipe for endless delays and disaster. The only way to avoid a nervous breakdown is to adopt the Congolese mentality—accept reality when things do not go your way and keep your sense of humor. Perhaps Congo and its people were teaching me life lessons that would make me a slightly better human.

It is difficult to express the satisfaction I felt when I awoke in a horizontal position in the Établissements St. Marc on the fifteenth of June. The hotel was in the usual cinderblock building with a corrugated roof, but the room was much larger, about one hundred square feet, and it had a window and a bathroom. Most luxurious of all, it had a single light bulb in the center of the room, and a power outlet to recharge batteries. The night before I had stayed up late to preserve the water snake and sample its DNA, a messy and stinky job given its state of decay. Not surprisingly, I had slept in, and when I opened my door to find my companions, I was blinded by the bright sunshine. Carrying a packet of instant coffee and a small can of Nido baby formula that I used for creamer, I wandered past two goats that were munching on trash around the hotel. I found Aristote and Wandege, beers in hand, sitting under a *perruque* in the courtyard behind our rooms. Not far from them, Poignard was cooking food over a small fire, and he moved frenetically between a small pile of ingredients and the pot. He seemed to be completely focused on his task, and he barely acknowledged me as I took a seat next to Wandege.

"How did you pazz de nightuh?" he asked.

"I slept like the dead," I responded.

He knew enough English to look surprised by my response. "You wazz dead?" he asked, half-jokingly, with an amused grin.

Aristote laughed and explained that it was a figure of speech.

I looked at their half-drunken bottles of beer. "*Pombe* (alcohol) before breakfast, eh?" I remarked, emphasizing the Swahili word for alcohol.

Wandege smiled with a sheepish expression, but unashamed, Aristote responded, "Yes!" He leaned forward in his chair and pointed a finger at the sky to ensure he would have my attention. "Yesterday was hard, it was VERY hard, and so this morning, yes, why not have two *pombe*?" he asked. He relaxed back into his chair and crossed his arms, confident there could be no objection to his logic.

"Two?" I replied.

"Yes," he continued. "Do you see this bottle and the one by Wandege?"

I nodded as I glanced at the large bottles of Primus beer near them.

"Well, these bottles are the second for us today," he said with a satisfying grin.

"Oh!" I responded. "I guess *pombe* is your breakfast this morning!"

"Nooooo," Aristote replied with aggrievement. "Everybody is VERY hungry, and we must eat soon." His eyes glanced over to Poignard with a dubious look as he said, "Maybe the food will be ok."

I asked him if Chifundera was still asleep. Aristote told me that Chif, with his typical superhuman stamina, had left the hotel hours before to look for a dugout canoe for the next leg of our expedition. I wanted to go to Lake Mai-Ndombe, meaning "dark water," for the unusual color of the lake. It was Congo's largest lake in the western part of the country, only a few herpetologists had ever set foot there, and it seemed isolated enough that it might contain endemic species—animals that can be found in one place and nowhere else. I wondered whether it might have an unknown species of turtle that had escaped notice, especially since herpetologists tend to focus more on terrestrial animals. There were also rumors of a strange water cobra that looked a bit different from the more widespread and common species.

Kwamouth is situated at the confluence of the Congo and Kwa rivers, and with a motorized boat, I planned to travel about sixty miles east along the Kwa, join the Fimi River at a town called Mushie, and continue another one hundred miles to the east until we reached the southern edge of the lake. With plenty of time to explore, we could stop anywhere along the way if someplace looked particularly promising. But I was also curious to see what animals we could find, if any, in and around Kwamouth. It had the same mixture of vegetation

I had observed since Maluku—tall reeds near the bank, followed by small hills with exposed soil and rocks, grass, palm trees, and clumps of short shrubs.

Before my imagined journey to Mai Ndombe could run away with me, I saw Poignard approaching me with a steaming cup of purified water for my coffee and a plate. A deep pang in my belly reminded me that I had not eaten much since we left Maluku, and I held my breath as I waited to see what he had prepared. He mumbled something unintelligible in Swahili as he gingerly placed the cup, plate, and a fork in front of me. Expecting the quintessential Congolese staples of rice and beans, I was pleasantly surprised to see a perfectly prepared vegetable omelet, fried plantains, a small banana, and fresh pineapple chunks.

"Wow!" I exclaimed. "A for presentation. And where in the world did he get *eggs*?"

"There is a small market in the town," Aristote explained. "Never mind. How does it taste?"

I realized I had been selected to be the guinea pig for our new cook. I plunged my fork into the omelet, admiring the consistency of the eggs, and took a bite. "It's good!" I said. Next I tried the plantains and was really impressed, because he had used some kind of secret spice to jazz up the flavor. "Wow, this is the best breakfast I have ever had in Congo!" I did not think Poignard could understand my English, but my positive reaction was universal, and he smiled with pride. Aristote translated my compliment, and Poignard nodded his head enthusiastically.

"Eyyyyy," said Aristote as he pointed at my plate, as if to say, where is mine?

Poignard replied, "*Nakuja*" (I am coming) in Swahili as he returned to his fire to bring breakfast for Aristote and Wandege. By the time we had finished a few minutes later, we reached a consensus that Poignard was a talented cook and his probation was over. When Chifundera returned later and tasted the food, he immediately agreed with us.

To make his position official, I paid my first installment to Poignard, and later that afternoon, he returned from the market with new jet-black sunglasses. Perhaps he wanted to fit in with us, because we had sunglasses to protect our eyes from the bright sun on the river. His new accessory also augmented the self-confident swagger he displayed everywhere he went. But there was something untoward about the way the large and opaque lenses completely covered his eyes in a manner that seemed to form a partial mask of his face. From that moment on, whenever he donned his sunglasses, it was impossible to gauge his mood.

In the evening, I joined Chifundera, Wandege, and Aristote to look for frogs on the southwestern edge of town, where Chifundera had spotted an artificial fishpond that was likely to attract frogs. A cool breeze wafted up from the Congo as we approached, and I could hear the calls of several types of frogs emanating from reeds that had colonized the edge of the pond. In short order, we found four common species of rocket frogs (*Ptychadena*), toads (*Sclerophrys*), tiny puddle frogs (*Phrynobatrachus*), and spiny reed frogs (*Afrixalus*).

Attracted by our headlamps, legions of sand flies swarmed around us, and worried about their bites, I hastened to finish our work. Female sandflies of the genus *Phlebotomus* are capable of transmitting a nasty *Leishmania* parasite in the Old World, causing over a million new infections every year.[10] There are different forms of leishmaniasis disease, including a common form that causes skin sores and ulcers, and a relatively rare and deadly "visceral" form that affects internal organs, resulting in tens of thousands of deaths every year.[11] Because there are currently no vaccines or prophylactic medications available, the only viable defense is to avoid bites.[12] The neighboring countries of South Sudan and Uganda are part of an East African hotspot of the deadly visceral form of the disease, but there is a worrying lack of data for Congo, most likely because it does not have the resources to monitor this neglected tropical disease.[13] Although I was unsure whether the disease-carrying sand flies occurred along this stretch of the Congo, I did not want anyone on the team to end up as a case study. Out of an abundance of caution, we retreated back to the hotel, where we enjoyed a spectacular goat stew from our new cook.

Following another outstanding breakfast from Poignard on June 16, I was in my room with Aristote, both of us busy photographing the frogs we had captured at the fishpond. Chif had found a dugout canoe to transport us east along the Kwa, and I was eager to wrap up our work in Kwamouth so that we could depart the following day. Aristote paused his official duties as photographic frog wrangler long enough to answer a knock at the door with an inquisitive stare at the person outside.

"*Mbote*" (greetings), said a gravelly voice to Aristote. A middle-aged man cocked his head through the doorway, and when he saw me, he waved and said, "Good morning, Misteer Eli!"

It was rare to encounter a Congolese person who was fluent in English, even one with a thick accent, so I was intrigued. I set my camera down and shook the man's hand at the doorway. He was skinny, of average height, and his clothes seemed to be two sizes too big. With bloodshot eyes and a disarming smile, he explained that he was a Congolese journalist named Jimmy. He heard that a foreign biologist had arrived in town, and he decided to pay me a visit to tell me about the bonobos that were in a reserve a few hours north of us near the Congo River. One of the last large mammals to be discovered in 1929, the bonobo had been confused with the similar-looking chimpanzee for years until scientists working in a Belgian museum noticed a striking difference in their skull morphology. The bonobo is now recognized as a distinct species of great ape, although its closest relative is the chimpanzee, and it can only be found in rainforests south of the Congo River.[14]

"Your English is quite good," I remarked, as I tried to figure out if he was after a story or wanted to tell me one of his own.

He raised his head with pride and replied, "Thank you! I am fluent in English, Swahili, Lingala, French, German, Spanish . . . and I have a working knowledge of Italian!"

Aristote, fluent in four of these languages, eyed him suspiciously.

Speaking in a loud voice reminiscent of a zealous preacher, Jimmy continued talking about the bonobo reserve with such enthusiastic gestures that spittle started whizzing out of the corner of his mouth on to the floor. Aristote noticed and took a step back to avoid the spray. I was simultaneously amused and intrigued.

"Nkala!" he continued. "Nkala is the name of the place, only three hours north of here, and the Belgians are there! They are habituating the bonobos for tourists, and university students are studying them too."

"And there is forest there, where these bonobos live?" I asked.

"Yesssss!" Jimmy exclaimed. "Beautiful untouched forest, because Nkala is a reserve to protect the bonobos!"

Now he had my full attention. Pristine forest was becoming increasingly rare in Congo, mainly because of locals razing the land to grow crops and foreign logging companies extracting lucrative hardwoods like mahogany and ebony. One of the main goals of my National Science Foundation grant was to study the Congo River as a potential barrier to rainforest-endemic amphibians and reptiles, and a stretch of virgin forest near the river would be invaluable to collect data to test this hypothesis. But all of this sounded too good to be true, because we had not observed large forest patches near the river yet.

I told Jimmy that I had seen a recent *National Geographic* article about bonobos, and I was under the impression that Salonga National Park, represented by an enormous and remote pair of forests hundreds of miles to the northeast, was their main stronghold in western Congo.[15] He said that when the leading bonobo experts heard the story that I was hearing now about a decade before, they were skeptical too. But now everyone was so convinced that an international conservation organization had established a field station at Nkala to habituate the bonobos.

These bonobos were hidden from the outside world until a 2005 survey by an all-Congolese team, including Bila-Isia Inogwabini, who eventually became a professor and scientific director at the Jesuit Loyola University of Congo. Over the course of eight months, this indefatigable survey team passed through over 160 miles of rainforest and swamp forest to document several bonobo populations near Lakes Tumba and Mai-Ndombe. The largest populations, along with the largest group sizes, were recorded from the area in and near Nkala.[16] This finding was very surprising given the prevalence of non-forest habitats at the site, which are not typically associated with bonobos. Because of Nkala's savanna-forest mosaic landscape, Inogwabini and his colleagues hypothesized that the bonobos take advantage of savanna fruits when the forest ones are out of season. They also speculated that bonobos walk upright when traveling in the savanna habitat.[17] If true, this must be an astonishing sight reminiscent of our human ancestors, who also walked upright in savanna habitats to spot predators and view the landscape more efficiently.[18]

With a quiet knock on the room's half-open door, Chifundera and Wandege joined the conversation. Jimmy greeted them politely, but then quickly returned the focus to Nkala and the bonobos.

"How have the bonobos survived without protection?" he asked rhetorically. With a dramatic flourish, he reached into a worn briefcase and produced a notebook. When he opened it and everyone crowded near him to take a look, I was surprised to see that the writing was in English. Reading verbatim from his notes, Jimmy explained that the local Teke tribe believed the bonobo was descended from a great warrior who had become indebted to a man. To escape an ancient law that required the warrior to become the slave of his creditor, he had hidden in the forest. Now the bonobo was considered taboo, and the Teke never hunted them. He finished by assuring us we would only have to walk half a mile into the forest to see them.

Perspiration poured off Jimmy's wrinkled brow as he completed his sermon-like speech, and he closed his notebook with a snap. Chifundera was studying me carefully to observe my reaction. I smiled at him. The story, opportunity, and adventure were too beguiling to dismiss.

"Change of plans?" Chif asked.

"Yes," I said. "Tomorrow we will head north."

THE SHAMELESS APE

In evolutionary biology, the term *diversification* refers to the splitting of one lineage of organisms into two, thus increasing taxonomic diversity over time,[1] and this is the process by which most new vertebrate species are formed. Several million years ago, well before diversification events that led to humans (*Homo sapiens*), our ancestors and those of chimpanzees (*Pan troglodytes*) and bonobos (*Pan paniscus*) were one and the same. Scientists have debated whether this creature resembled a bonobo or a chimp more closely, but it was certainly chimpanzee-like.[2] Similarly to our closest living relatives, there is an excellent chance that this ape could climb well, and it probably avoided nocturnal predation by ascending trees in its equatorial forest home.[3] Back then, in the mid-Miocene epoch, there were impressive numbers of large carnivores in Africa, including fearsome amphicyonids, commonly known as bear-dogs. Based on fossils of one species (*Megamphicyon giganteus*) known from Namibia to Spain, scientists estimate this behemoth could weigh over 1,300 pounds, and it probably terrorized our ancestors before it went extinct.[4]

As the Miocene progressed, the planet experienced global cooling, and Africa's extensive rainforests receded, giving way to drier savannas.[5] These vegetation shifts were also caused by changes in atmospheric circulation and rainfall patterns, which resulted from the orogeny of mountains in eastern Africa.[6] No other continent changed as dramatically as Africa did at this time, and many of its plants and animals slipped into extinction in a widely recognized faunal turnover event.[7] In the middle of this dynamic change, the lineage that would give rise to the human species split off from the one that would lead to chimpanzees and bonobos. Different studies estimate that this diversification event happened about 6–10 million years ago, but most estimates are from 7–8 million years ago.[8] The lineage that led to the human genus *Homo* entered a brave new world in East Africa's savannas, where it evolved improved bipedalism (the ability to walk upright on two legs), a larger brain, and eventually, the ability to use fire and understand the text you are reading now.[9]

After the proto-human lineage parted ways with the apes, the ancestral lineage of bonobos/chimpanzees endured without changing much for several million years. Then in the early Pleistocene, about 1–2 million years ago, this lineage experienced diversification again, splitting into the modern-day chimpanzee and bonobo.[10] A group of Japanese researchers proposed the "conditional corridor hypothesis" as the most likely mechanism to explain the origin of bonobos—a small group of chimpanzee-like apes crossed the Congo River at one of its shallowest points in the northeastern section, perhaps because of a rare reduction in discharge linked to cyclical periods of aridity in the Pleistocene.[11] Because these apes are poor swimmers, the small "founder" population that managed to cross the Congo did not return, and over time, the genetic makeup of the founders changed to such a degree that they eventually became a different species, the bonobo. Although scientists have reached a consensus that the chimpanzee and bonobo are distinct species, recent genomic studies suggest that genetic admixture (exchange of genetic material between species via interbreeding) did occur on a limited scale after the diversification event.[12]

No matter how dark the nightclub, nobody would ever mistake any of these great apes for a human being.[13] But aside from their genetic differences, how are bonobos distinguished from chimpanzees? Although bonobos have also been called pygmy chimpanzees in the past, they are only slightly smaller than chimps, but bonobos tend to have a more gracile appearance, with a slender body, thin neck, narrow shoulders, a small head, small ears, and long legs. Bonobo faces are relatively dark and flat with high foreheads, the lips are reddish, and their long, black hair is always parted in the middle of their heads in a fashion similar to Alfalfa of the Little Rascals. Because of the raised position of their hips, bonobos have a straighter back when they are upright, giving them a humanoid appearance.[14]

The behavioral differences between chimpanzees and bonobos are stark. Chimps tend to have anger management issues, and their male-dominated societies include frequent bouts of violence, and sometimes even murder, both within and between groups. Wild chimps are known to kill each other in cruel ways, and in one extraordinary case in Uganda, six human children were abducted from a rural village and eviscerated. In contrast, matriarchal bonobos are less excitable and more sensitive, at times painfully delicate in their emotions. As an example, in World War II during a bombing raid in Germany, three bonobos at Munich's Hellabrunn Zoo seem to have died of fright (officially, heart failure), but the chimps in the cage next door survived. Other notable distinctions of the bonobo include larger groups, stronger female bonds, a tendency to be more vocal with high-pitched voices resembling squeaky toys, and a habit of shaking their hands when calling.[15] Bonobos are remarkably tolerant of other groups and frequently cooperate, share food, and form alliances with them.[16]

The most famous difference between chimps and bonobos is undoubtedly their sex life. As one might expect from the macho chimpanzees, males frequently use coercion, infanticide, or violence to induce mating with females, and they perform sex in only one position (male entering female from behind), because the vulva is oriented posteriorly. Bonobo females have their vulva positioned between their legs, allowing them to mate in a face-to-face position and many others, and violence is rarely a prerequisite to sex, perhaps because strong female-female bonds disarm unruly males. In fact, the bonobo sexual repertoire is kinky enough to make a porn star blush, and they are considered to be the most sexually active nonhuman primate on the planet. Bonobos are capable of having heterosexual or homosexual sex multiple times per day and in a diverse array of copulatory positions, often while mothers are holding their babies. They are known to engage in mouth-to-mouth kissing, and they can gratify genitals with their hands or mouths. Males have even engaged in penis-fencing games. Because the bonobos involved in this sexual contact sometimes change their facial expressions, make grasping movements, and even scream, researchers believe they have orgasms. Remarkably, bonobos begin all of this sexual activity before they are even a year old.[17]

An intriguing hypothesis by Harvard primatologist Richard Wrangham explains the origin of the peaceful nature of bonobos, and its key component is food.[18] Chimpanzees and gorillas have substantial overlap in their diets, and recently documented attacks by chimps on outnumbered gorillas in Gabon occurred during periods of food scarcity.[19] Although evidence is sparse (gorillas lacked a fossil record until the twenty-first century), it is conceivable that a gorilla-like ape once lived in the forests on the left bank of the Congo River, where bonobos are endemic today. A cold and dry period that occurred about two million years ago could have temporarily killed off perennial herbs, the main food source for gorillas, thus wiping them out everywhere in Congo save for the Albertine Rift Mountains east of the right bank. When the climate bounced back, and with their competition removed and food abundant in all seasons, the bonobo had no reason to go to war, and less aggressive behavior evolved over time.[20]

Considered together, bonobos and chimpanzees offer unique glimpses into human nature and evolution. On the one hand, it is easy to understand the link between violent behavior in chimpanzees and humans, as well as their competitive, male-dominated societies. Because proximate groups of chimps are usually engaged in different degrees of hostility with one another, one could argue that they are intolerant and xenophobic too. But, on the other hand, bonobos offer mirrored lessons about the origins of humanity in *Homo sapiens*, including altruism, tolerance, empathy, and living in peace with one's neighbors. Echoes of the primordial human soul, both good and evil, can be traced back to these apes.

We still have much to learn about chimps, bonobos, and ourselves, but research on wild populations should be hastened, because the growing human footprint (e.g., population density, infrastructure development, presence of roads), poaching, habitat destruction, and climate change are reducing great ape numbers all over the world.[21] Chimpanzees have a large distribution from the forests and savannas of West Africa to Tanzania, but many of their populations are patchy and decreasing, and they are currently listed as an endangered species.[22]

The bonobo's distribution is less patchy, but it is restricted to rainforest on the left bank of the Congo River, its population estimates are decreasing, and it, too, is listed as an endangered species.[23] If we let our closest relatives disappear, several chapters of ancient human prehistory will never be read and questions about our natural origins will remain a mystery forever.

On the afternoon of June 17, after the usual frustrating delays, I found myself in a dugout canoe speeding away from Kwamouth. I sat behind Jimmy as he pointed out notable landmarks on our journey north. Countless clumps of reeds and water hyacinth drifted south, occasionally with hitchhiking birds that probed the vegetation in search of a meal. We passed by several dugouts, some only slightly larger than iron boards, with perfectly balanced fishermen standing on each end. As we passed near one of them, I greeted a fisherman with a friendly "Mbote" and he relaxed his grip on his spatula-shaped oar long enough to wave back.

Aristote waved too, adding the standard Lingala reply "Mbote, sango nini?" (Hi, what is the news?) He smiled at me, delighted to have an opportunity to teach me more of the language. When he turned his head back, he spotted a strange-looking building near the bank and asked, "What do you think that is—maybe from the Belgians?"

When I looked closer, it seemed to be a shipping container that had been converted into a small building, complete with cutout openings for a door and windows. A large hut was nearby, and in the distance, I could see additional buildings, including one made of bricks and corrugated iron that seemed to be on the verge of collapse. Two young boys wearing oversized shorts stared at us from the reeds near the bank. Jimmy followed our gaze and jumped in with a complex answer.

"No, this is not from the Belgians," he began. "It was a rabbit farm of the former president Mobutu. They were killing elephants here, and there was also . . . eh, what can I say . . . a mobile clinic here."

"They were killing elephants here?" I asked in shock.

"Elephants, yes!" Jimmy confirmed. "Chinese people were growing crops here too."

I was surprised to hear that elephants were hunted in this area during Mobutu's heyday in the late twentieth century, not because the habitat was unsuitable for them, but because it was so close to the Congo River. Starting in the late nineteenth century, Belgian colonizers used the river as a highway to penetrate the Congo Basin (about 1.3 million square miles of lowland rainforest, the second largest contiguous tropical forest in the world[24]), and they probably shot every elephant in sight to obtain their valuable ivory.[25] The commodity was exported to Europe to make tooth replacements, billiard balls, piano keys, combs, knife handles, snuff boxes, and carved curios.[26] Some have argued that ivory exploitation was a crucial step to formation of the colonial Congo Free State (1885–1908), during which an estimated thirteen thousand elephants were slaughtered every year.[27] No wonder that in Conrad's famous novella *Heart of Darkness*, the protagonist Marlow remarks, "The word 'ivory' rang in the air, was whispered, was sighed. You would think they were praying to it."[28]

Whether Jimmy's story was true or an exaggeration, there was no doubt that the elephants near the Congo River were long gone now. If the Belgians had not done them in, surely they were wiped out later, either by poachers in the reign of Mobutu or soldiers during the war. Any traumatized survivors would have learned, long ago, to avoid people at all costs.

We continued north in silence for hours, and I enjoyed the tranquil scenery, but Jimmy's estimate of a three-hour journey was a bit off. After five hours we entered a narrow stretch of the Congo in complete darkness, and some distance ahead I could see twinkling golden lights through the palm trees near the bank. But as we drew closer, I realized that the lights were on the western side, in Republic of the Congo. Jimmy told me that the town of Mpouya in the other Congo had good restaurants, modern hotels, well-stocked shops, and several bars. Civilization tempted us, but we dared not venture into another country at night, and my gaze turned to our destination on the relatively dark and quiet eastern side.

The town of Tshumbiri had been the site of a Protestant mission in the colonial era,[29] but now it seemed to be a small collection of crumbling buildings and huts. There was no trace of any electricity, because I could see only LED lights and flickering candles in a market that was still open for business. With no time to make a fire and cook food, we ate stale bread for dinner, and I washed it down with a grenadine soda from the market. With no hotel on the eastern side of the river, we stamped down vegetation as high as our waists and pitched our tents in a grassy field on the edge of town. For hours, I tried in vain to fall asleep while an army of crickets chirped in the grass around us, but eventually exhaustion worked its magic.

Chifundera roused me at dawn with bad news. All our bread had run out, the market was closed, and we were about seven miles west of Nkala. Of course, nobody in Tshumbiri had a truck, and it seemed there was no way to reach our destination. At least Poignard had found wood for a fire to make everybody tea or coffee, but it would take a while to cook rice and beans, the Congolese staples for most of our meals. Irked that we had been led into a dead end, I found Jimmy talking to a woman near the edge of our camp.

When he saw the look on my face, he tried to calm me with a warm smile. "My friend, how did you sleep? My rheumatism is not good for me today."

I cut to the chase. "Not bad, Jimmy, and sorry to hear about your rheumatism, but now we seem to be stuck here. Any ideas?"

"Yes, of course, no problem," he began. "Just a small distance away, we can reach Nkala after a few minutes . . ." He started to stutter as he realized I was not buying his rosy conception of the distance. "Uh uh, my friend . . . do not worry, I think we can find a motorcycle to transport someone to Nkala, and then he can ask the people there for a truck to bring the others . . . if, if they still have one."

"You mean beg for a truck as our only hope, Jimmy?" I said with thinly veiled sarcasm.

"Well, yes, of course," Jimmy replied with an overcompensating smile.

I sighed and looked at Chifundera, who was wide-eyed with astonishment. He gave me a sarcastic grin, shrugged his shoulders, and said, "Ok, I will look for a moto."

In the late morning, Chifundera returned with a teenage boy on a dirt bike. Chif said that I should go so that he could look for some food for our dinner, which we all hoped to eat in Nkala. I scrambled onto the black cushioned seat of the moto behind the boy, winced slightly from the sunbaked heat radiating from it, and held on for dear life. He hit the gas, and we took off like a rocket. We passed through miles of narrow footpaths that were flanked by bushes, trees, and parched elephant grass, all of which smacked my face and arms as we weaved through it. Several times our progress was delayed as we struggled over hills so steep that we had to get off the bike so that we could push it to level ground. A few trees cast shadows of shade on our path, but much of the ground had been baking in the sun all morning, and the heat emanating from the savanna was intense.

When we reached Nkala over an hour and a half later, I noticed several well-maintained buildings, and a group of people sitting in a circle under the shade of an acacia tree. The person leading the discussion was a white woman, and she waved at me as I arrived in a cloud of dust and sweat. I paid off my driver and stood in the shade of the closest building while I waited politely for the meeting to end. When I looked to the east, I could see beautiful rainforest in the distance, and I grew excited to be in a place where no herpetologist had ever set foot before.

The gathering finished and a relatively tall German woman introduced herself as Inga, the project manager for the conservation program in the area. Very friendly and welcoming, she had an easygoing manner that was disarming, and it seemed that my luck was changing. Indeed, I had been fortunate to catch her on her weekly visit to Nkala. We chatted for a while about her collaboration with local trackers to habituate the bonobos for future tourists. They had made impressive progress in recent months, but the meeting had been about them losing track of the apes, and days of searching had failed to relocate them. Because bonobos often hang out in the canopy of the forest, it is easy for them to avoid detection if they wish, and this talent likely saved many of the ones in Salonga National Park from being wiped out during the war.[30]

When I explained my situation, Inga immediately agreed to send a truck to fetch the team and our gear in exchange for a modest amount of fuel. Once the ball was in motion for this to happen, she invited me into the nearest building to escape the heat. I was impressed by the new paint, modern furniture, and especially the cleanliness of everything. I took a seat at a small dining table, where I was introduced to Martine, a young Belgian student, and Phillipe, a Congolese biologist. When I noticed everyone's crisp and clean clothes, and compared them with my sweat-soaked attire, I felt a bit embarrassed by my grungy appearance.

The conversation bounced between French and English as we all explained how we had come to Nkala. Before her current focus with bonobos, Inga had worked on other projects, including tiger conservation in Asia. Martine had spent over a year in Congo studying bonobo ecology along with other Belgian students. Phillipe was one of the Congolese staff hired to support the project. When I talked about herpetology and showed them a makeshift field guide I had put together for the trip, Phillipe pointed to a softshell turtle (*Cycloderma aubryi*) photo, licked his lips, and exclaimed, "Ah, je connais celui-ci. Vous pouvez le manger!" (Ah, I know this one. You can eat it). Inga laughed and said that he made the same comment about many different types of animals. We chatted for an hour or two, and I wished everyone luck with their bonobo work, even if the stars of the show were nowhere to be seen at the moment.

Eventually, the rest of the team showed up in the truck, and I introduced everyone to Inga, Martine, and Phillipe. Inga allowed us to set up our field camp at the edge of the forest next to some vacant buildings that seemed to be made out of branches, mud, and corrugated metal coverings. Because we had not eaten much, Poignard darted away to begin his frenzied search for firewood so that he could start cooking our dinner. I had been so busy setting up the camp that I had not noticed the cloth bag dangling from Wandege's hand until he waved it in front of me with a smile. When he opened it, I was astonished to see the contents.

Somehow he had managed to capture not one but two sand snakes (*Psammophis lineatus*). These lightning-fast snakes are active during the day in grassy habitats, where they chase down lizards and envenomate them with fangs located at the rear of their mouth. One of the more bizarre adaptations of these snakes is the evolution of relatively small and smooth hemipenes (male snakes have two "intromittent" organs for copulation) to avoid predation while mating. The theory is that these snakes are easy to spot during the day in open habitats, and if a predator shows up, the male can disengage quickly to ensure a quick escape for both lovers. They are also one of the few known snakes that have the ability to twist their body in a way that can deliberately break off their tail if it is seized by a predator. In fact, one of the snakes was missing a tail, providing evidence of a narrow escape in the past. Although sand snake venom is not deadly, it can be painful, and my thoughts immediately returned to Wandege's near-death experience with the stiletto snake. But he smiled and assured me he had not been bitten during the captures.[31]

"Kazi njema!" (Good job!) I exclaimed in his native Swahili.

I barely had a moment to celebrate Wandege's success before Aristote showed up next to a man holding a stick with a dead snake dangling from it. "A farmer in the next village killed this snake by his house this morning. When he heard we were looking for snakes, he brought it to us," Aristote said.

Like almost any place in the world, most people in Congo are fearful of snakes, and understandably, they do not want them to harm their livestock, pets, or family. Unfortunately, it is impossible for most people to recognize whether a snake is venomous or harmless, and so they are usually killed on sight. As long as the body has not decomposed much, it is still invaluable as a museum specimen and DNA sample for research. When I glanced at the small snake, I saw a blunt, marbled grayish head with small eyes and catlike pupils. Its body had an attractive pattern of dark brown chevrons and blotches on a brown and tawny background. It was an egg-eating snake (*Dasypeltis confusa*), named for its bewildering variation in color pattern that had caused taxonomic confusion for years.[32] Aristote and I interviewed the farmer to record the details of where and how the snake was spotted, and when we were finished, I thanked him for his time.

African egg-eating snakes are truly a marvel of adaptation to a specialized food source. These snakes have a reduced number of tiny teeth that allow them to engulf bird eggs several times wider than their heads. Bones of the skull and jaw have been modified to allow the mouth to expand in this way, and elastic skin and accordion-like folds of soft tissue assist the process. The snake's mouth and neck resemble a balloon as the egg passes down the throat. Eventually, the fragile food is squeezed against a series of vertebrae with thickened spines that act like can openers to slice open the shell and prevent it from moving too far into the back of the body. The egg's contents pass further into the digestive tract, but the empty shell is regurgitated. Although egg-eating snakes are completely harmless, many of them resemble the color patterns of dangerous vipers that live in the same area. Remarkably, some species can even mimic the warning sound of deadly saw-scaled vipers (*Echis*), which rub their serrated scales together to create a rasping sound similar to water sizzling on a hot plate. If potential predators can be fooled by the egg-eating snake's ruse, I could not blame the farmer for playing it safe to dispatch the snake near his house.[33]

Shortly after arriving in Nkala, Jimmy had disappeared, but Chifundera found a local man to serve as a guide, and we invited him to dinner. We celebrated the promising luck at Nkala with rice and chicken, complete with *pili pili* (hot chiles) that add spice to any Congolese meal. All of us passed our rave reviews of the food along to Poignard, who nodded modestly in agreement. Satiated and content, we relaxed and talked as the sun went down, listening to the crackling fire and growing chorus of night-shift insects. In the skies above the glare of the fire, the orange glow of sunset cast an eerie backlight to the forest, giving it a dark and foreboding appearance. For a moment, nobody said anything as we seemed to absorb the presence of the wilderness around us.

I broke the silence by asking, "Who is ready to find out what animals are inside that forest?"

Chif, Wandege, and Aristote responded enthusiastically, and we retreated to our tents to gather our gear for nocturnal frog hunting. I offered everyone mosquito spray, but in unison, they refused because of the noxious smell. Feeling a surge of adrenaline, I followed behind Chif on a narrow, meandering path that led into the depths of the forest. Our guide told us that it had not rained for weeks, and although the vegetation seemed to be dry, a thick wall of humidity enveloped our bodies as we entered the new realm, and I started sweating immediately.

The packed density of trees, shrubs, and lianas closed in around us, giving the impression that we had entered a city of vegetation. Veritable skyscrapers were composed of 150-foot-tall *Gilbertiodendron* and *Entandrophragma* trees with thick crowns of leaves that blocked any view of the stars. I was reminded of the words of Cuthbert Christy, a British explorer who truly captured the essence of a Congolese rainforest when he said, "The height and size of the forest giants and the magnitude of the buttresses of some of the leviathans are astonishing, and everywhere beneath the vast canopy, as within a great cathedral, is a dim light, a weird stillness and an awe-inspiring silence."[34]

Whenever we wandered off the trail to flip a log in search of sheltering reptiles, our boots made rustling noises as they swished through a carpet of dead leaves. Occasionally, these sounds were punctuated by louder crunching sounds as we stepped on fine roots or twigs buried beneath the leaf litter. At times one had to duck beneath overhanging branches and lianas, or scamper over half-rotten trees that had died after toppling over the trail. We did our best to avoid the impressive thorns of palm fronds, but it was impossible to avoid snagging one's sleeve against them while passing through the thickest masses of foliage.

Off in the distance, we heard something calling *"Hoo"* in a humanlike voice that seemed to echo between the great trees before going silent. We all stopped in our tracks.

"Chif, was that a bonobo?" I asked in disbelief.

"I . . . I don't know," he replied honestly. Our guide said that it was a bonobo.

For a moment, nobody spoke or moved as we waited to hear if the call would repeat itself, but it never did. I will forever wonder whether the sound we heard was from an ape or some other animal. But we had entered Nkala's forest in search of other creatures, and we soon moved on.

Our guide led us closer to a stream, and as we approached it, I started hearing the telltale quack-like call of forest treefrogs (*Leptopelis*). I climbed into a small tree above the stream to catch a lime-green frog that was perched on a high branch, while my colleagues picked up a handful of tan and brown ones on shrubs near the water. A few hours of searching turned up several additional kinds of frogs, including white-lipped (*Hylarana*), rocket (*Ptychadena*), reed (*Hyperolius*), and spiny reed frogs (*Afrixalus*), but nothing that seemed to be rare or new. As we returned to our tents for the night, we made plans with our guide to penetrate deeper into the forest next time, because it was possible we were too close to the edge to encounter the rarest of the rare.

Reverberating calls of a territorial great blue turaco (*Corythaeola cristata*) roused me at dawn. Resembling a muppet gargling through a megaphone, "*Gahhhrook!*" was screamed repeatedly, followed by more rapidly frequent "*Prru!*" sounds that diminish in volume to a kind of bubbling purr.[35] One of the larger and more spectacular birds of Central African forests, this species is immediately recognizable by its vibrant azure plumage, including a long tail, yellow beak with orangey red, lipstick-like accents at its tip, and a bluish-black crest of feathers in the shape of a mohawk. Greenish parts of the belly and tail are attributed to a copper-rich pigment called turacoverdin, which is only found in African turacos.[36] The showy feathers and long tailfeathers are more flamboyant than functional, because the birds prefer to glide or leap from tree to tree with assistance from semi-zygodactyl toes that can grip branches like a vice while it searches for food, including fruit, berries, leaves, and flowers. Over the course of five to thirteen days, mating pairs build large nests, where the female lays one to three nearly spherical, blue-green eggs that hatch after about a month or so of incubation by both parents.[37] Given the bird's enormous range in equatorial Africa, it is not currently threatened with extinction.[38]

I emerged from my tent to find Poignard wandering around camp with his winter coat zipped up tightly around him. Somebody in the village was already busy at work chopping wood, and I heard some kids playing nearby. The sound of my tent zipper was loud enough to alert Aristote that I was awake, and he leaned his head out from behind one of the rudimentary buildings to look at me.

"Can you come see this insect?" he asked.

Curious, I wandered around the corner and saw two boys prodding an enormous cricket with a stick. I leaned in closer to take a look at it. At least two inches long, the giant African armored cricket (Tettigoniidae[39]) had a shield-like structure on its back that was bordered by several sharp spines. The body was brown, the legs were cream with brown spots, and its grayish antennae probed the ground as it crawled along. The boys were using the stick to avoid bites from the insect while also hoping to elicit its most spectacular defense strategy. When attacked, these crickets and several other types of insects are known to deploy an autohemorrhage reaction, whereby nasty smelling hemolymph (i.e., insect blood) is reflexively ejected up to a foot away from the creases of the leg joints and from under the pronotum, the large protective plate on its back. Experiments have shown that lizards will readily attack the crickets, but once the autohemorrhage defense occurs, the would-be predators become distressed and immediately spit them out.[40] I made a mental note to leave these insects in peace if I ever came across one.

Another Belgian student wandered over to see what the commotion was about, and I chatted with him about his research on bonobo foraging. Martine and Inga joined us too. We were engaged in a lively conversation about the challenges of bonobo research when we heard a woman raising her voice behind us. An extremely thin, middle-aged woman in a red *kitenge* was angrily accosting a young man near her, who was looking away nonchalantly in an attempt to ignore her.

"Oh dear, it's Gloria," said Inga.

When the woman noticed me looking at her, she seemed to forget her anger and smiled at me. I returned the smile and waved, but realized that something was not quite right with her.

"I told her so many times, you must go to Kinshasa, you are sick," Inga said with a sad voice. "She used to be such a nice woman, but now she gets angry so easily, and I fear she is starting to go mad."

Martine chimed in with her thick French accent by saying, "Yesss, she must meet zeh doctor in Kinshasa or zeh sleeping sickness will kill her."

I had been a little sleepy during the conversation because I did not have my morning cup of coffee just yet, but the mention of sleeping sickness jolted me to attention. Known in the medical establishment as Human African Trypanosomiasis (HAT), few tropical diseases in Africa are more horrifying. In Central and West Africa, the affliction is caused by a subspecies of worm-shaped, unicellular parasite called *Trypanosoma brucei gambiense*, with the genus name derived from the Greek *trypaô* (boring) and *soma* (body) for its corkscrew-like movements.[41] The parasite is introduced to its human host via the bite of a tsetse fly (*Glossina*) vector, which has a similar size and appearance to a horse fly. Once inside a human body, the parasite is devastating. Because it is enveloped in a continually modified glycoprotein that can evade the immune system, infections become chronic and impossible to purge without treatment. In the first stage of their assault, the parasites reproduce in the blood and lymphatic system, but eventually, they breach the blood-brain barrier and attack the central nervous system. Infections can progress over the course of years, sometimes up to fifteen years, causing a bewildering array of symptoms including disrupted sleep patterns, neurological problems, and psychiatric episodes before the inevitable body wasting, coma, and death.[42] Gloria seemed to be suffering from the advanced stage of the disease.

The terrible toll of HAT and similar trypanosomes that infected other animals likely shaped the history of Africa by making certain kinds of agriculture, transportation, settlements, and development of civilizations with centralized political structures impossible in areas with high densities of tsetse flies.[43] When domestic horses arrived in West Africa, several kingdoms developed calvary, a game-changer for wars, but trypanosomes carried by tsetse flies blocked the adoption of horses in many other areas of Africa.[44] Sleeping sickness was almost eliminated in the 1960s, but thanks to wars and a collapse of surveillance programs in postcolonial Africa, the disease returned with a vengeance in the late twentieth century, exceeding over 300,000 cases a year.[45] For half a century, the drug of choice to cure the disease was melarsoprol, a compound based on arsenic and dissolved in diluted propylene glycol, an ingredient for car antifreeze.[46] Patients described a feeling of hot chili peppers in their veins during the ten-day course of painful injections. Not surprisingly, the drug had harsh side effects and even killed around 5 percent of people who received it.[47] Thankfully, increased funding from the Bill and Melinda Gates Foundation, among others, led to the development of less draconian drugs to treat the disease, including acoziborole, which seems to have the potential to cure the disease with a single dose.[48] Pan-African tsetse fly eradication and HAT-monitoring efforts have been highly effective, because the case numbers in recent years (mostly in Congo) have dipped below 1,000.[49] A study published in 2022 noted optimistically that "HAT is closer than ever to being eliminated."[50] Nevertheless, as Gloria proved, the disease was still alive and well in Nkala and many other areas of western Congo.

I looked at my European acquaintances with concern and asked, "Aren't you afraid you might get this too?"

Martine dismissed my concern with a wave of her hand, saying "The flies move very swowly, and you can just reach out wiz your hand and crush zem like zis!" She squashed an imaginary, lone fly against a tree with a slow push of her wrist, as if she were extinguishing the insect equivalent of a sloth. My skeptical brain immediately conjured Ernest Hemingway's *Green Hills of Africa*, in which he described "everywhere were tse-tse flies, swarming around you, biting hard on your neck, through your shirt, on arms, and behind the ears."[51] Hemingway's tormentors seemed like darting ninjas, not the clumsy and slow ones that Martine described. I hoped I would never have firsthand knowledge of the flies to reconcile the disparate accounts of their abilities.

Curious, I asked them if they were taking precautions for malaria. Wanting to avoid a repeat of my seven previous cases in Congo, I had been taking daily prophylaxis pills since I left Texas, but they had annoying side effects, including fatigue. They told me that they did not bother, because they were unaware of any cases in the area. I had trouble masking my shocked reaction. Their contention was hard to believe given the huge numbers of infections in some parts of Kinshasa,[52] and even higher case counts in rural areas of western Congo.[53] However, I thought they knew the area well, they had managed to avoid the disease for many months, and perhaps they were right. Nonetheless, I continued to take my medication. I am glad I did.

I set up a laboratory in one of the mud buildings, and Aristote and I got to work photographing the frogs we had found the previous night. Poignard looked on with an odd stare as he peeled potatoes with a knife. Hours later when the work was complete, Chifundera appeared to say that the men who brought us to Tshumbiri in their boat were demanding more money. When I asked why, he said that they could not find passengers to return to Kwamouth, and we needed to pay for them to go back. If we refused to pay, they might go to the local police and create a problem. Frustrated but wiser, I handed over the money, along with extra funds for his transportation to Tshumbiri.

"I think you should add a little extra to buy beer for the chief," Chifundera added. He had contacted the local leader to inform him of our activities, and the man wanted to meet me. I agreed to Chifundera's request, and he bought a few Primus beers as presents in Tshumbiri. After his return in the late afternoon, all of us walked to the chief's outpost near Nkala. When we arrived, no women were in sight, but all of the village's men and boys seemed to be encircling the chief, who was seated on a wooden chair. The gaunt, middle-aged man was dressed in a bright red jumpsuit, accessorized by a faux-gold watch. He grasped a scepter that was topped with buffalo hair, a ceremonial symbol of his power. We accepted seats near the chief, Chifundera explained our mission, a few cups of beer were shared, and the chief gave his blessing for our work to continue.

Before delving deeper into Nkala's forests that evening, Aristote fetched a ski mask from his tent and pulled it over his head. The effect was comical, but perhaps he had a practical purpose.

"Aristote, is that mask to protect you from mosquito bites?" I asked.

"Yes, maybe," he replied with a shrug.

I laughed and said, "Well, I don't know how well it will protect you from mosquitoes, but maybe it will bring you good luck."

In fact, it did. That night we encountered more of the same species we had the night before, but we also found an unknown species of fingernail-sized squeaker frog (*Arthroleptis*) with a brown back, cream flanks, orangey legs, and a white spot on its belly. They were hopping through vegetation on the forest floor like insects, and we noticed that we could tap the leaves with a stick to induce the tiny amphibians to jump out into view.[54] In the moist leaf litter along the edge of a small stream, Wandege found the most spectacular frog of the night. Slightly less than an inch long, the frog had a bronze iris, a light brown back with irregular black spots, a black "mask" bordered by a thin white line that extended from its nose to the middle of its flanks, and a honeycomb pattern of black blotches separated by anastomosing light gray lines that transitioned into aqua blue on its hind legs.

I immediately recognized Wandege's discovery as a new species of frog, and subsequent DNA analyses would confirm my hunch. In 2015, my colleagues and I named it the Congolian long-fingered frog (*Cardioglossa congolia*), based on this specimen and only one other male that would be found the following day.[55] Long-fingered frogs get their name from the ridiculously enlarged third finger in males of most species, a unique character that they share with their closely related cousins, the squeaker frogs (*Arthroleptis*).[56] The exact purpose of this elongated and often spine-fringed digit (up to 40 percent of the body length) remained a mystery until recently.[57] A squeaker frog encounter first reported by French herpetologist Jean-Louis Amiet in 1989 offered an intriguing clue: "These frogs were standing on their hind limbs, with the fore limbs of one under the arms of the other, and with their elongate third fingers vibrating against each other[']s backs."[58] In 2020, scientists actually videotaped a nearly identical male-male combat fight in another species of squeaker frog,[59] but this behavior has not been seen in long-fingered frogs, at least not yet. Many long-fingered frog species are known from few specimens, suggesting they are either rare, difficult to find, or both.[60] Although additional examples of *Cardioglossa congolia* were recently found by Congolese herpetologists at another bonobo reserve hundreds of miles northeast of Nkala, females remain unknown.[61]

Over the next few days, it rained sporadically, alerting me to a leak in the ceiling of my tent, but the moisture also drew out more animals that had been hiding during the dry spell. Aristote continued to wear his ski mask during our nightly excursions into the forest, and when he found several rare frogs, I remarked that his mask must be magic. He agreed and continued wearing the good luck charm for the rest of our time at Nkala. We encountered a trove of additional snakes, lizards, and frogs from savanna and rainforest habitats, and it was very exciting to see such an incredible diversity of life in one place.

One afternoon, Chifundera and Aristote returned from the forest with an unexpected creature. While searching under the bark of a dead tree where lizards sometimes take shelter, they found a grayish-blue scorpion with a yellowish tail and limbs. I instinctively recoiled when he showed me a plastic container, and I saw the spiderlike legs of the invertebrate scurrying across the bottom of it in a way that reminded me of the movie *Alien*.

And yet I was also fascinated. If scorpions survived four mass extinction events over the course of their 400-million-year history on Earth,[62] including the one that wiped out the dinosaurs, they can surely teach us something about survival. Terrestrial scorpions like the one before me evolved from aquatic ancestors that probably crawled on to land as air-breathing arachnids about 325–350 million years ago, slightly later than the salamander-like vertebrates that made their first tentative forays on to land in the late Devonian period.[63] A recent tally estimated there are over 2,300 species of scorpions worldwide, with more than 1,000 of these described in just the last 20 years, but many more new species likely await discovery.[64]

"Yikes," I said. "Well, our target isn't really scorpions, but I suppose we can sample it because it might be rare. How bad do you think the venom is?"

Chifundera agreed that it might be rare and added, "The venom is as bad as an *Atractaspis*." Once again, my mind returned to Wandege's near-fatal bite. I was especially skittish as we photographed the scorpion, being very careful to stay far away from its potentially deadly stinger.

Later that year, an expert would identify the specimen as *Uroplectes occidentalis*, a relatively common buthid scorpion in Congo and Angola.[65] Aside from a few studies that are focused on taxonomy, little is known about this particular species, and there is a dearth of knowledge about Central African scorpions in general.[66] A related southern African species (*U. vittatus*) stings many people who collect firewood, and although very painful, the venom is not known to be deadly.[67] Nonetheless, Chifundera was right to think that we should be careful, because our poorly studied scorpion might be dangerously venomous.

The Buthidae family of scorpions is the largest and most widespread group of scorpions, with more than one thousand species, including some of the world's deadliest. The infamous Egyptian deathstalker scorpion (*Leiurus quinquestriatus*) is one of the most venomous species in the world, with neurotoxins similar to those of some deadly snakes.[68] I made a mental note to be more careful when flipping logs in the jungle as I searched for sheltering snakes, because a nastier surprise might be lurking underneath.

As our time at Nkala drew to a close, the bonobos remained elusive, but we had found so many rare and new species of amphibians and reptiles that I could only think optimistically. Pleased with the team's hard work, I promised everyone that I would buy them some beer as a reward when we passed through the next town. We laughed as Aristote danced around the evening fire in anticipatory celebration of the *pombe*. Even Poignard, usually stoic, cracked a smile. Little did we know that a deadly invader was already reproducing in two of us.

LAIR OF THE WATER COBRA

I made a deal with Inga to catch an overland ride from Nkala to a town called Nioki about eighty-five miles to the east on the Fimi River. The Soeurs de l'Immaculée Conception (Sisters of the Immaculate Conception) hotel had a generator for power, there was an airstrip nearby, and I kept my promise to buy everyone some beer with our dinner at a modest restaurant. Given the relatively developed state of the town, I had high hopes that it would be easy to find a boat to take us east to Kutu, near the southern edge of Lake Mai-Ndombe. From there, we planned to head north into the heart of the lake to explore its swampy forests.

But transportation in remote areas of Congo is easier said than done. One does not simply saunter down to the river, strike up a conversation with a boat captain, and hire them to go where you wish. No, you must find the actual owner of the boat and negotiate with them, followed by another person who owns an outboard motor, then a captain, and then the fuel. Of course, the downside of towns and cities in Congo is DGM, which requires special permits for river transportation, especially for a foreigner like me. This process took most of the day, but we managed to leave in the afternoon without the drama we had experienced at Maluku.

The straight-line distance from Nioki to Kutu is about thirty miles, and the captain assured us we would reach it in a few hours. We were nearly there, but when it started to get dark, he insisted we spend the night at a tiny fishing village with a handful of huts on a sandbank of the river. In the morning we were eager to move on to reach Kutu, but a heated conversation between Chifundera and the captain portended trouble.

"Le carburant est fini!" (The fuel is finished!) the captain roared at Chifundera, as if it were his fault.

Actually, the fuel was not completely gone yet, but we definitely did not have enough to reach Kutu. Somehow the captain and engine owner claimed they did not know how far we wanted to go, even though we were very clear our first stop was Kutu, the next sizable town as one heads east along the Fimi from Nioki. Perhaps the motor had consumed more than anticipated, but whatever the reason, we seemed to be in trouble. We argued with them for a while, until they agreed to continue east as far as our fuel would take us. After all, what else could we do? We had not passed anything for miles, so going back was not an option.

We did not travel far before the motor began to sputter, but we managed to reach an old Belgian logging port called Kempili Bembe before we completely ran out of fuel. It must have been abandoned decades ago because everything was crumbling away, including a small warehouse, rusting boats, and decrepit cranes. As we neared the bank, I spied the mangled remnants of steel tracks that must have been used with heavy machinery to load timber on to boats headed for Kinshasa in a bygone era.

The captain and engine owner stormed off with a jerry can in search of fuel, and a small group of children appeared to stare at us. While we waited around for two hours, the children helped us chase down some lizards that were basking on the concrete remains of the lumber yard. A speckle-lipped mabuya turned up, along with four lizards that looked like their faces had been decorated with bits of a peach. They had orangey snouts featuring a bright yellow streak that ran from the lower jaw past the neck along the side of the body. With thick scales arranged in transverse rows that resemble a suit of medieval plate armor, these so-called African plated lizards (*Gerrhosaurus*) are distinguished by their reinforced scales, composed of bony osteoderms, along with a peculiar fold along each side of the body.[1] The function of these armor-like scales is probably defensive, at least from the bites of other individuals during fights, but many alternative hypotheses ranging from thermoregulation to camouflage remain untested.[2] I ran my thumb along the body of the most colorful male, and it reminded me of a braided leather belt.

Eventually our boat companions returned with a small amount of fuel to reach Kutu, which was only three miles away. We had traveled about half of this distance when Wandege spotted a dead snake hanging from vegetation near the bank of the river. We paused our eastward travel long enough to retrieve the emerald green serpent, which proved to be an Angolan green snake (*Philothamnus angolensis*).[3] These harmless snakes average two to three feet long, and they spend their days climbing through waterside vegetation as they hunt for frogs and nestling birds.[4] Exactly how this individual met its demise was not clear, but I speculated it was killed and accidentally dropped by a bird of prey.

As we continued east along a particularly pretty stretch of riverside forest, hundreds of white butterflies flittered above the water. It was still early when we arrived in Kutu and the entrance to the lake was not far away, so Chifundera left with the captain to look for more fuel. There were only a few scattered trees here and there, and most of the town was surrounded by tall grass and reeds near the river. The bleak landscape suggested Kutu would be exceptionally hot by mid-afternoon. Three hours later, tired and sweaty from the relentless sun, Chifundera returned to say that all the fuel in Kutu was gone. Having no choice, we moved into a group of dusty storage sheds that somebody was renting as hotel rooms near the port, and before we could really settle in, a renewed argument with the captain and engine owner commenced. Eager to get a viable DNA sample from the green snake before it decomposed further, I ducked into my room to set up a laboratory and process the specimen.

The afternoon sun had warmed the windowless room to a veritable blast furnace, so I left the door open, and I could hear the argument continue for the better part of an hour. I was covered in sweat by the time my work was completed, and I was debating whether I would escape the heat to observe the argument when Aristote appeared at the door.

"Can you come see another *muzungu* (white man) here?" he asked.

"A white man, *here*?" I asked in disbelief.

"Yessss," Aristote replied, seemingly unsurprised.

Given the skyrocketing temperature and humid air, the dispute had shifted to the shade under a *perruque* in the courtyard of the hotel. The sun did not discourage a sizable crowd of spectators from gawking on the sidelines, including Poignard, who was somehow donning his winter coat. In the middle of the group of gesticulating men, I noticed a middle-aged white man, who had apparently injected himself into the middle of the argument in perfect Lingala! I had heard his voice while working in my room, but I had not detected a Belgian accent until I listened more carefully. Stocky, tanned, disheveled, and drunk, he seemed to be reveling in the quarrel, but I could not tell which side he was on. When he and the crowd noticed me, the conversation died down, because the meeting of two white men in Kutu was probably a very rare occurrence. Everybody watched me as I approached the man.

His hand was muscular and leathery as he greeted me with a crooked smile. "Je suis Patrick" (I am Patrick), he said in French.

I talked with him for a few minutes, but he seemed cagey when I started asking innocent questions, including the obvious one about how the hell he ended up in Kutu. He rubbed his fingers along the edge of his beer with one hand and cupped his stubbly chin with the other as he told me he was in the fish business, and that he traveled back and forth to Kinshasa to "faire des affaires" (do business). When I paused to contemplate how he could transfer fish all the way to Kinshasa before spoiling, he noticed my look of puzzlement. He shot another crooked smile at me, tacitly acknowledging that his fish tale was bullshit, and then he changed the subject to ask about me.

When I told him I was looking for reptiles, he jumped into a story about a recent snake encounter that was so ebullient I needed Aristote to help me translate his avalanche of French. I ducked into my room to grab my field guide, and when I showed him the section on snakes, he pointed at several photos, exclaiming "Oui, je connais celle ci!" (Yes, I know this one!). This seemed to be even more entertaining than the argument, and the crowd of spectators closed in around us to look at the photos. Based on their reactions, I did not need to speak Lingala to understand that many of them knew the snakes as well. Word spread quickly, and soon the entire town knew that we were looking for reptiles.

I retreated to the "hotel" in an effort to escape the heat, but it was in vain. Even as the sun was starting to set and cracking sounds from the corrugated iron roof suggested it was cooling, the inferno in my room lingered. Trickles of sweat zigzagged down my back as I constantly sipped water to replace my fluids. I tried to do some reading, but my focus wandered as if I were dreaming. Poignard whipped up spicy chicken and fries, but I could eat only a little, because my stomach revolted when I attempted to add more heat to my body. Dizziness and exhaustion crept into me.

The air around Chifundera seemed to shimmer as he entered the room with bad news. He, too, was sweating, and based on the exhausted look on his face, I could tell that he had lost the argument. The boat captain was creating more problems than he was worth, and Chif recommended paying him even more money to cancel our agreement and get rid of him. Too lethargic to question his logic, I handed over the money and hoped for the best. I knew we were now effectively stranded in Kutu, but *c'est la vie* in Congo.

I felt slightly better than comatose when Aristote burst into my room at dawn to tell me that a boy had found a really interesting lizard at a place called Ibia on Lake Mai-Ndombe. Half-awake, I stumbled out of bed, slowly realizing that the heat was less intense, at least temporarily. I took a couple deep swigs of water, swatted away some mosquitoes, and squeezed my feet into my jungle boots. Evaporation from my sweat-soaked clothes was slowed by the humid air, but modest as it was, it made me feel cooler as I started walking.

Aristote was standing next to a dour-looking waif who was clutching a large bucket. I smiled at him and said, "Ok, voyons voir" (Let's see). He looked at me blankly, not understanding my French. Aristote said something to him in Lingala, took the bucket, and with a dramatic "Voilà!" removed the lid to show me the contents. If I had been sleepy before, I was now wide awake.

Inside was a two-foot-long lizard with a long tail, sharp claws, and a beautiful pattern of yellow bands and spots on a dark gray body. It was a young Nile monitor (*Varanus niloticus*), Africa's most geographically widespread lizard, occurring in areas near freshwater from Egypt all the way to South Africa.[5] Feeding on everything from freshwater crabs to crocodile eggs, these versatile predators can grow over eight feet long as adults. They use their ribbon-shaped whip of a tail to swim surprisingly well, sometimes staying submerged for twenty minutes, and they can even corral fish into shallow water where they can be eaten more easily.[6] They are exceptionally wary and quick, and when I had seen them in the past, it was usually as a blur of scales as they leaped from tree branches into water and out of sight.

Over the next couple of days, we worked with the locals to find more animals, including geckos, sand snakes, water snakes, plated lizards, skinks, and frogs. In the afternoons, we spent most of our time hiding from the relentless sun as we tried to find a way to continue our journey to the lake. At last on the twenty-eighth of June, we left Kutu at dawn in another dugout canoe and headed east toward Lake Mai-Ndombe. Gradually, the endless reeds, tall grass, and patches of forest that had dominated the scenery since Nioka gave way to an increasing number of rainforest trees near the bank.

Mai-Ndombe is a tree-shaped lake that is approximately 25 miles long, 37 miles wide at its northern edge, tapering to 6 miles wide in the south, with an average depth of only 10 feet, but it fluctuates by about 6 feet with the rainy and dry seasons.[7] American[8] ichthyologists working at the lake in the 1970s noticed that when they put their hands just one foot deep into the acidic and dark brown water, they could not see them anymore.[9] One might assume that the harsh chemical properties of the water, including the low dissolved oxygen content, would discourage aquatic life. And yet there is a surprisingly diverse fish fauna, including the endangered and endemic cichlid *Nanochromis transvestitus*, so named for the reversed sexual dichromatism of the species, whereby females are brightly colored and males are drab.[10]

As we turned north into the lake, the color of the water transitioned from the dull brown typical of most rivers in Congo to exceptionally dark brown water that gave the lake its name.[11] The edge of it was now dominated by large rocks with overhanging trees, including white "bathtub ring" deposits on the upper parts of the rocks from the fluctuating water levels. In some places, the lake was so wide that I could not make out the bank on the other side. As the bow of the boat cut through small waves formed by a steady wind and light rain, foam-like bubbles frothed around it, giving the impression that we were traveling through an ocean of root beer. The temperature dropped, and everybody scrambled to dig out their jackets, except Poignard, who was already shrouded in his winter coat as usual.

The English missionary George Grenfell, considered to be the third greatest Congo explorer after Stanley and David Livingstone,[12] entered the lake on his steamer *Peace* in 1886. He saw a few crocodiles, and numerous buffalo drinking at the shore, and the boat was repeatedly attacked by territorial hippos.[13] As we traveled deeper into the lake, the shores seemed lifeless, except for a few birds. We passed by a relatively small logging operation and a couple of fishing villages on the southern edge of the lake, but as we continued north for many hours, most of the forest looked intact, and there were few villages visible on the shore. The rain produced a cloudy mist on the surface of the water and nobody said a word—the wind and constant hum of the outboard motor were the only sounds. In one particularly thick patch of mist, Poignard glanced back at me from the bow, presumably to observe how I was dealing with the cold. I was actually savoring the cool weather, but I caught an odd expression on his face that was difficult to peg. Was it guilt or fear? I could not shake the eerie feeling that we were traveling among ghosts, and indeed, this was not far from the truth.

Long ago, the area surrounding Mai-Ndombe had been ruled by the Bolia kingdom and several chiefdoms with similar political systems. The Bolia were fierce warriors and thirty-eight successions of kings were recorded from their court traditions.[14] In the late nineteenth century when Congo was carved into different concessions for commercial exploitation of its natural resources, the lake (known then as Lac Leopold II) became part of King Leopold's Domaine de la Couronne (Crown Estate).[15] Taking advantage of soaring demand for rubber after the invention of the inflatable tire in 1888, the regional governor set up rubber quotas as a way of taxing Africans living within the king's territory.[16] Profits exceeded 700 percent and greed inevitably led to atrocities. Enforcement by the notorious Force Publique was brutal beyond words, and included imprisonment, rape, amputation, and murder. One reputable estimate stated that about half of Congo's population was wiped out between 1880 and 1920.[17]

In the late afternoon, we landed on the eastern shore near Inongo, the largest town in the area, which included an airport, bank, prison, and schools.[18] Chifundera asked around and we decided to set up our camp at a tiny fishing village called Mpote-Emange, population circa one hundred, at the edge of seemingly untouched forest. It was close enough to the lake that we did not have to carry our luggage far, but not too far from Inongo to find transportation for the next leg of the expedition. I was also hoping that there might be a chance to work with the fishermen to catch some of the lake's aquatic reptiles, including water cobras. When I entered the small clearing where the village was located, I saw several rudimentary buildings that seemed to be constructed out of mud bricks and large branches, most with thatched roofs.

Nearly all of the people who watched our entrance were in their teens or younger, but as I reached a shady spot that Chifundera had selected for our campsite, I noticed him talking to an ancient barefooted man who was using a stick as a cane. When the man noticed me, he exclaimed "*Ooh la la!*" to express his surprise at seeing a white man. I smiled and introduced myself to Mputu, the chief of Mpote-Emange, and wondered how his black cargo shorts and red Cleveland Indians T-shirt had found their way to him.[19] Once Chif had explained why we had come, he graciously welcomed us to his village and assured us that many animals lived in the nearby forests.

As if to underscore the chief's point, I looked up and noticed a huge hornbill bird perched in one of the trees above us, seemingly eavesdropping on our conversation. Based on its black feathers and yellow bill with a reddish-black tip, it was an African pied hornbill (*Lophoceros fasciatus*), a relatively common species in lowland rainforests of equatorial Africa. In Congo, it seems to prefer clearings and secondary forest (damaged by human activities and with a different composition and structure from nearby primary or "virgin" forests), which explains how I saw it in Mpote-Emange.[20] Like other hornbills, eggs are laid in natural cavities of trees, sometimes over 120 feet above the ground. The female uses her own droppings to seal herself inside the cavity, except for a narrow slit to receive food from the male while she incubates the eggs. They eat mostly fruit and insects, suggesting that the village had fruiting trees nearby.[21]

Wandege and Aristote were already setting up their tent near the edge of the forest and I followed suit, glancing up briefly at the one-hundred-foot-tall giant tree next to me, now covered in shadow from the waning sun. By the time I was finished, darkness had prompted a symphony of insect calls from the forest, beckoning us to explore them. Chifundera was feeling tired so he turned in early, but Aristote, Wandege, and I plunged into the foliage to see what we could find. In stark contrast to the dense terra firma forest we had seen at Nkala, many of the *Uapaca* trees visible to us in the swamp forest of Mpote-Emange were more widely dispersed and had either widened buttresses or so-called stilt roots at their bases. Stilt roots resemble curvy branches that emerge near the base of the tree, often at awkward angles, before turning downward into the soil. They are especially common in swamp forests and mangroves, where they probably serve as adaptations to soft, waterlogged soils by improving stability and oxygen supplies. Amazingly, some stilt roots may facilitate young palm trees to "walk" around obstacles such as rocks by permitting them to lean sideways, send out new roots, and discard the lower trunk and older roots, thus allowing the crown to move to a new location.[22]

Based on the presence of large amounts of decomposing leaf litter on the ground, it was likely that our visit coincided with a recent flooding event, because trees in swamp forests will typically shed their leaves when the roots are inundated with water.[23] I surmised that during the wet season, the forest probably floods to such a degree that it merges with the lake water, perhaps contributing the decaying organic matter that makes the water so dark. Now that we were in the dry season, most of the ground was relatively dry, but pools of water persisted in shallow depressions around us. We had to watch our steps carefully as we searched the forest. Aristote wisely carried a stick to probe the depth of puddles we encountered, because sometimes, floating leaf litter on the water's surface deceived us into thinking it was shallow when it was not. Some places had larger, stagnant ponds that contained a diverse array of aquatic invertebrates (including plenty of mosquitoes), freshwater fish, and frogs.

Within the jumble of leaf litter, decomposing logs, mud, and scattered understory plants, dozens of fingernail-sized squeaker frogs (*Arthroleptis*) hopped to and fro, their marbled pattern of brown, gray, and black blending in perfectly with the vegetation and earth. Light brown forest toads (*Sclerophrys*) also hopped along the forest floor, their spiny tubercles resembling a bad case of chicken pox all over their back and legs. An orangish brown green-eyed frog (*Cryptothylax greshoffii*) turned up near the edge of one of the pools of water. Undoubtedly the stars of the evening were the golden puddle frogs (*Phrynobatrachus auritus*), named for their yellowish coloration, at least in some African populations. The individuals we found were a beautiful mix of brown and orange, with a dramatic black mask (with minute whitish flecks) that extended from the tip of the snout to the forearm. Just as we grew tired and were about to call it quits, Wandege noticed a strange *Hemidactylus* gecko on the trunk of a tree.[24] It was covered in small cream tubercles, including its eyelids, and it seemed to be looking down on a white-lipped frog (*Hylarana albolabris*) that was perched on a leaf just underneath it.

Delighted with our bounty, we headed back to camp, but as we drew near, I started to hear a strange call emanating from the crown of one of the tallest trees. The sound was a frog-like "*Gonk!*" repeated in rapid succession, but the pitch did not seem quite right, and the volume was so powerful that it echoed. Could a frog really be calling from a one-hundred-foot perch? We stopped in our tracks to listen more carefully, and after a pause, Aristote solved the mystery.

"Bats," he said flatly. "They will call all night until they get tired."

Sure enough, even though my tent was about a hundred feet away from the tree where they were calling, and despite my attempt to muffle the noise with earplugs, the monotone screams kept us awake until they got tired at about four a.m. The calls can be described as either a wheezing clown horn, or an amplified and endless game of ping pong that is repeated with rapid, maddening succession. During the sleepless night, I wondered what the purpose of the calls could be. One possibility was uncovered by a group of Israeli scientists. Using a machine-learning algorithm, they studied vocalizations of the Egyptian fruit bat (*Rousettus aegyptiacus*) and suggested that they squabbled over food (sometimes with their mouths full), their sleeping position, unwanted sexual advances, and personal space. Incredibly, they could even modify their calls when "speaking" to different individuals in a similar way to the tone of our voice when addressing a child or an adult.[25] These algorithms are being used to try to decipher communications in other species ranging from birds to whales, and in the not-too-distant future, the prospect of some human-animal communication might become a reality.[26]

On subsequent nights, we did catch glimpses of the animals in flight at dusk and confirmed that they were Old World fruit bats (Pteropodidae), also known as flying foxes. With their doglike faces, these bats resemble Chihuahuas with wings, both in approximate size, appearance, and even behavior. We tried to shine our strongest headlamp beams into the tree where the calls originated, but our view was obscured by branches and leaves, and we never came even close to identifying them as one of the fourteen genera or twenty-eight species known from Africa.[27]

However, I managed to record the call of the bats, and years later, when I compared them with archived recordings at the Macaulay Library at Cornell University,[28] I had a lucky break and found a perfect match.[29] With a face only a mother could love, the hammer-headed fruit bat (*Hypsignathus monstrosus*) is Africa's largest fruit bat, and it is infamous for the racket it creates while calling for mates.[30] It is one of only a few bats in the world that are known to engage in lekking behavior, whereby males perform on a stage of sorts (with no resources like food),[31] allowing females to select the singer she likes best for mating.[32] I surmised that the bats ceased calling in the wee hours of the morning because they either grew exhausted or got lucky.

Feeding on fruit, flowers, and flower products, Old World fruit bats are an integral part of the forest ecosystem, because they exclusively pollinate or disperse the seeds of some trees. The flowers of these trees have evolved to open at night, luring the bats with their irresistible fragrance.[33] If bat populations decline or disappear, so can the trees, creating a negative "ripple effect" in the forest for many other organisms that have ecological interactions with these trees.[34] As an increasing human population comes into contact with bats more frequently, we tempt fate, because many of the viruses infecting bats can potentially jump to humans in "zoonotic spillover" events, in which pathogens jump from wild animals to humans.[35] Such spillover events have happened frequently in the past, because the majority (more than 60 percent) of human infectious diseases are descended from pathogens that originally infected nonhuman animals.[36] The species of hammer-headed fruit bats I heard in Mpote-Emange have been identified as a likely reservoir (the human, animal, or environment in which infectious organisms grow and multiply in nature[37]) of the deadly Ebola virus.[38] The evidence linking this species of fruit bat and one other one (Franquet's epauletted fruit bat, *Epomops franqueti*) to Ebola was especially strong in a 2008 outbreak in Luebo, Congo, that killed 186 out of 264 infected people (70 percent). But despite the danger, many people in the Congo Basin continue exposing themselves by hunting fruit bats as an easily obtainable source of protein, especially as larger game species become scarce from habitat destruction and overhunting.[39]

But as we all know from the recent COVID-19 pandemic, Ebola is not the only virus that can potentially jump from bats to humans and wreak havoc.[40] An eerily prescient study from 2018 drives home this point with its title, "Bats, Coronaviruses, and Deforestation: Toward the Emergence of Novel Infectious Diseases?" Because Southeast Asia has razed about 30 percent of its forests over the last forty years, many species of bats are losing their habitats, leaving them a choice as old as life itself—adapt or die. Many species have taken advantage of human "anthropized" environments to survive, including houses and barns for cave-dwelling bats, and orchards and fields for fruit bats. This unnatural living arrangement attracts a diverse group of bats and their viruses to areas near humans living in poverty with concomitant unsanitary circumstances, creating perfect conditions for new pandemics to emerge.[41] Moreover, climate change will force bats and other animals to migrate to new areas, creating opportunities for an estimated four thousand spillover events in the next half century.[42]

Considering the timeframe from 1950 to the first decade of the twenty-first century, deforestation rates in the Congo Basin were not nearly as devastating as those in the Amazon or Asia,[43] but this is likely to change in the coming years. Africa's human population is exploding, and a quarter of the world's people will be African by 2050.[44] Several kinds of coronaviruses, including some novel ones, have already been detected in African bats living near human settlements, including those from the Congo Basin.[45] In February 2025, the World Health Organization (WHO) announced fifty-three people had died from a mystery illness in Congo, not far from Mbandaka, after three children ate a bat carcass and kicked off a spillover event.[46]

Many other African viruses should be on our radar too. A more lethal form of Mpox, formerly known as monkeypox, is another zoonotic virus (likely linked to rodents and nonhuman primates) that has been increasing in prevalence in Congo recently.[47] In August 2024, the WHO declared Mpox an international public health emergency for the second time in two years as cases spread to Burundi, Kenya, Rwanda, Uganda, and Switzerland.[48] Predictably, by November 2024, the first case in the United States was found in California.[49] Central African bushmeat hunters who are HIV-positive and immunocompromised are especially worrisome because they are ill-equipped to fight off infections from nonhuman primates that they capture, butcher, and eat, sometimes after being bitten or scratched. Under these circumstances, it is relatively easy for a variety of nasty viruses, including Mpox, to jump from animals to humans.[50]

There is no way to predict when or from where the next pandemic will pop up, but there is a good chance it will originate from Africa. As the famous Latin saying goes, *ex Africa semper aliquid novi*, meaning there is always something new out of Africa. But just because something is new does not mean it is good.

Soon after emerging from my tent on the morning of June 29, I decided to take a walk alone to see if I could get lucky and find some rare reptile basking in the slivers of sunlight that dappled the forest floor. I was struck by the airy spaces between trees and the huge amounts of leaf litter filling them, because in other Congolese forests, these spaces would be crammed with many more trees and understory vegetation. The stilt roots of some trees were so elaborate that they formed elaborate latticeworks. When the echoing call of a great blue turaco broke the silence, it was easy to imagine that I was trespassing into a mythical woodland. I flipped several logs, probed tree holes with my snake tongs (sometimes doubling as a reacher-grabber tool), and searched branches above my head for arboreal snakes, but all in vain.

Knowing that I had many hours of photographing and laboratory work ahead of me, I headed back to the village, stopping to behold the colossal tree where the hammer-headed fruit bats had serenaded us the previous night. It was a one-hundred-foot-tall fig tree[51] with bark that had the tan color and bumpy knots of a ginger root. Its sprawling branches merged with those of equally tall, adjacent trees to form a beautiful matrix containing hues of green and yellow. Lianas resembling angel-hair pasta hung from some of the branches, and I spied some kind of bird nest in one of the uppermost tree forks.

With over eight hundred species occurring in the tropics and subtropics across the globe, *Ficus* is one of the largest plant genera in the world, and they comprise about half of the diversity of the mulberry family (Moraceae).[52] Only a fraction of figs are actually trees—the rest are climbers, shrubs, herbs, rheophytes (aquatic plants), lithophytes (plants that grow on bare rocks), epiphytes (plants that grow on other plants but are not parasitic), or hemi-epiphytes, which establish themselves in the canopy of rainforest trees and send down roots or shoots into the ground far beneath them.[53] Although there are exceptions, many species of fig are pollinated by their own specialist species of fig wasp, and these two groups of organisms have coevolved together since the heyday of dinosaurs about seventy-five million years ago.[54] Some of these tree species produce an exceptional bounty of ripe figs that are rich in calcium throughout the year, potentially helping many kinds of fruit-eating animals to survive seasons when other fruits are scarce. This long list of animals includes hundreds of species of birds and mammals, and even some reptiles and fish. This unique biology has prompted some ecologists to label figs as keystone species, because they are essential to maintain the structure and integrity of living communities, including the survival of large networks of species that depend on them.[55]

Chifundera smiled at me when he saw me return from the forest, but he had dark circles around his eyes. He was clutching a cup of coffee and sat hunched over in a way that was unnatural for him. It was obvious that bad news was on the horizon. He told me that yet again the captain of the boat that had brought us here wanted double the money to return to Kutu. This time, Chif did not have the strength to argue with him. I handed over the money without protest, now realizing that return funds were the cost of dugout canoe transportation in Congo, even if this was not made clear in the initial negotiations.

Following hours of fruitless searching for animals, I was taking a rest to read a book in the shade when Aristote cleared his throat and slowly approached. Poignard eyed him from his cooking fire in a *perruque* at the edge of our camp. I knew something was up.

"Hello man, I have been talking to some fishermen, and they say the water cobra is here," he began.

"Really?" I asked as I raised my eyebrows in surprise.

"Yes, of course! They say they catch them regularly in their nets when they fish near mangroves, and if they find one, they will send someone for us to take it."

I heard a zipper moving and saw Chifundera moving out of the one-man tent he had insisted on bringing along for the entire expedition. It was barely enough room to contain his body, and he wiggled out of it, caterpillar-style, until he could get to his feet. He staggered a little as he did so. Aristote and I stopped talking and were staring at him, but he did not look up to greet us.

"Chif, are you ok?" I asked.

"Ahhhhh, it is malaria," he said in a hoarse voice. "I can feel the stiff headache originating from my upper neck, so this must be it. Every time it is the same. Wandege has it too."

Aristote cursed and I got up from my chair to offer assistance. Chifundera reassured us that he and Wandege were already taking good antimalarial pills that we had picked up in Kinshasa, and that they were used to it and would be fine. But whenever anyone in Congo gets malaria, it is cause for great concern, because Africa harbors *Plasmodium falciparum*, the most deadly species of the parasite that causes the disease. About 90 percent of the hundreds of thousands of annual deaths from malaria are reported from Africa, and Congo is among the worst-affected countries.[56]

I knew from my experience with seven previous malaria infections (or so I was told by rudimentary blood tests in poorly equipped clinics in Congo) that Wandege and Chif were facing a nasty ordeal. After the malaria parasites (rod-shaped sporozoites) enter the human body via the salivary glands of female *Anopheles* mosquitoes, they migrate to the liver where they replicate for the initial, asymptomatic incubation. Eventually, these initial invaders give rise to about forty thousand merozoites that rupture the liver cells, spill out into the blood stream, and begin the erythrocyte (blood cell) infection stage. The parasites undergo asexual reproduction and consume the cell's hemoglobin (the protein that carries oxygen in the blood) before it bursts, releasing more merozoites to infect more red blood cells. This is when all hell breaks loose in the human body. Jonathan Kingdon, who grew up in East Africa, described the symptoms as "an alternately sweating, aching furnace of limbs, torso and head turns to frozen jelly, shaking uncontrollably to become the inert and breakable shell around a loose and frightened spirit."[57] During the recurring waves of blood cell infection and bursting, one experiences a rollercoaster of fever paroxysms, muscle aches, fatigue, and headaches, followed by chills that can shake the entire body, even when covered by several warm blankets. Other unpleasant symptoms include jaundice, stomach pain, vomiting, diarrhea, and respiratory distress. The act of walking across a room feels like running a marathon. If the infection develops into severe malaria, death can result from organ failure.[58]

I paused for a moment to think about the incubation time of malaria. With rare exceptions, most people experience the onset of symptoms six to fourteen days after infection, although previously infected people with at least some immunity tend to develop symptoms toward the slightly later end of this range.[59] Two weeks before we had been at Kwamouth, then briefly at Tshumbiri, but most of the likely range of infection overlapped with Nkala. So much for the Europeans' contention that malaria was not a concern there.

We did what we could to make Chif and Wandege as comfortable as possible and then checked in with Poignard. As usual, he was under a *perruque* and preparing food by a smoky fire. He was also holding court with at least a dozen curious teenagers around him. When we explained the situation and asked him how he was feeling, he smiled with his macho bravado, waved his hand dismissively, and told us that he was fine. He was obviously trying to look tough in front of his admirers, but I noticed that Aristote's gaze lingered on him a few moments longer than expected. Poignard noticed too.

As we walked back toward our tents, Aristote was uncharacteristically silent until we were out of earshot of Poignard and his fan club. He grasped my wrist and pulled me a few feet into the forest where nobody would see or hear us.

In a quiet voice he said, "I don't think Poignard is who he says he is." His eyes were wide and he looked troubled.

"What do you mean?" I asked.

"Do you remember when we first met Poignard on the boat at Maluku?"

"Yes, I can't forget," I said honestly.

"Well, he was talking Mashi with Chifundera, saying that he is from his tribe."

"Yes, I remember," I said.

"But Chifundera told me after that he spoke Mashi strangely, like it wasn't his native language. And now when he was speaking with those childrens, he sounded like a man from Rwanda. Maybe because he was having fun, he forgot to hide his accent."

I considered what Aristote was telling me, and then asked, "Why do you think he would lie?"

"I don't know," Aristote replied. "But maybe it is because he did something bad. We must watch him carefully. He and I . . . we are becoming friends, and I will try to find out more when we are taking *pombe* together and his tongue is loose."

"Ok," I replied. "I think that is a good plan. But, of course, this means you will need money for *pombe* right?"

He smiled bashfully to acknowledge that his clever plan included funding for beer. "Mayyyybe you can give just a little money for us to have some *pombe* when Chif and Wandege are better. We are doing a *kazi njema* (good job), yes?"

"Yes, you are Aristote, but sometimes I think you are too smart for your own good. But this malaria is a big problem. How are you feeling? Any fever or headache?" I asked.

"Everything is ok, I never get malaria," he replied.

"Never?" I asked.

"No, never," he replied with a shrug of his shoulders. His body gesture was dismissive, as if the terrible disease could not affect him because he was too tough for it.

There are many possible genetic variants, known and unknown, that might explain Aristote's seeming invincibility to malaria.[60] The one with the strongest known protective effect is a textbook example of genetics that I teach to my university students every semester. Because humans are diploid, meaning we have two sets of chromosomes (one from mom and one from dad), we have two copies (i.e., alleles) of each one of our genes. Genes are the functional units of DNA that encode information to make polypeptides, which are chains of amino acids that form proteins (in part or whole). One of our genes encodes information to make a β (beta) polypeptide for the oxygen-carrying hemoglobin protein in blood. If both allele copies are genetically identical and normal, the individual is homozygous (symbolized by $\beta^A\beta^A$) for the gene. If the allele copies are different because one of them has at least one mutation (a heritable change in the DNA sequence of the gene), they are heterozygous (symbolized by $\beta^A\beta^S$). People who inherit two mutant copies of the gene from their parents (symbolized by $\beta^S\beta^S$) are homozygous, but because both allele copies are defective, abnormal folding of the hemoglobin protein and malformed red blood cells result. These misshapen cells are fragile and break easily, and they tend to clog in capillaries, leading to a debilitating and potentially deadly disease called sickle-cell anemia.[61]

Obviously, nobody wants to end up with this painful disease, but there is a flip side to it—the malaria parasite cannot reproduce in sickle cells. Heterozygotes ($\beta^A\beta^S$) produce both normal and defective hemoglobin, a phenomenon known as sickle-cell trait, and when deprived of oxygen (e.g., high-intensity exercise), they might experience some of the milder symptoms of sickle-cell anemia. In places where malaria is endemic, homozygotes have a disadvantage, because either sickle-cell anemia or malaria can kill them. These dual pressures create an advantage for heterozygotes (known as overdominance in genetics), and increase the odds that neither allele is purged from the population. Aristote might be a heterozygote, because he was resistant to malaria, and he did not suffer from the nerve or organ damage that is typical for sickle-cell anemia patients.[62]

That night I felt vulnerable as I lay awake listening to the bat calls. I worried about the team and considered my own malaria risk. Because the mutant form of the sickle-cell allele is strongly associated with people who live in malaria-endemic parts of the world, or are descended from ancestors who lived in them,[63] there was little chance that my European ancestry conferred any genetic resistance to the disease. All I could do was hope that my daily antimalaria prophylaxis would continue to protect me, even if it had failed several times in the past. If my luck ran out, at least I had very good medication from Kinshasa as a backup, or so I hoped. A recent study of medications obtained in Congo and Cameroon showed a surprisingly high percentage of substandard and falsified (i.e., counterfeit) drugs are sold in these countries.[64]

The next day, Wandege and Chif showed signs of improvement. I was reassured by their rapid recovery and grateful that the medication we purchased in Kinshasa seemed to be highly effective. I encouraged them to take it easy, but by the evening, Wandege was feeling well enough to join Aristote and me in the forest. In an effort to search for tadpoles, I brought along a net, and we dredged the bottom of one particularly large pond. When we focused our headlamps on the muddle of decaying leaf litter, sticks, and tarlike mud, we could see aquatic insects and tiny grayish frogs wriggling within it. Our fingers got covered in the gooey mud as we sifted through the mess to isolate the frogs, which turned out to be aquatic dwarf clawed frogs (*Hymenochirus curtipes*).[65] The genus is known for its skin glands that release the frog equivalent of cologne (i.e., chemosignals) to attract female mates, and at least one species has an unusual mating ritual that involves a series of underwater summersaults.[66] These frogs are included in the ancient Pipidae family (dating back to the Cretaceous), which includes forty-one species in Panama, South America, and sub-Saharan Africa.[67] My fingers stroked the granular skin of the pear-shaped frogs, and I wondered how the flipper-like hind feet were strong enough to propel them through the thick mud.

A handful of common snakes wandered into the village the next day, creating brief excitement during the otherwise fruitless search for animals. During the downtime, I cleaned my water purifier, which had been repeatedly clogged by the sediment and particulate matter of the lake water. As the days passed by, I grew antsy because the animal we had come to find, the banded water cobra (*Naja annulata*), remained elusive. We had already put the word out with local fishermen, who assured us that they regularly catch the snakes in their nets near mangroves at the edge of the lake. But these snakes can hide very well in rock formations at the shore or in holes in banks or trees, sometimes living there for long stretches of time before venturing out to hunt for fish.[68]

Like other closely related cobras, these snakes have a potent neurotoxic venom that is likely deadly to humans.[69] No detailed case studies of bites are known,[70] but one study of global snakebite mortality from the 1950s claimed the species is "one of the most dangerous reptiles in Gabon."[71] During a visit to Lake Tanganyika in the early 1930s, Harvard herpetologist Arthur Loveridge noted that "there is a widespread superstition prevalent among the fishermen that if a man is bitten [by a water cobra] in the water he should remain there until treatment in the shape of a water weed is brought to him."[72] Interesting, but definitely not an idea worth testing.

I was especially keen to find at least one of these water cobras in Lake Mai-Ndombe, because they seemed to look quite different from the typical form. A German book about venomous snakes of Africa from 2007 had shown photos of a captive specimen of water cobra from Mai-Ndombe, and its odd coloration was strikingly different from photos of the more widespread banded water cobra.[73] Reaching a total length of six feet or more, the typical coloration in Central Africa is yellowish to light brown with 21–23 distinct black rings around the body, most of which completely encircle the body.[74] The photos from the German book showed a grayish background color with white and cream spots on the flanks, and it seemed to lack bands altogether. Perhaps the Mai-Ndombe cobra was a new species, but I had to obtain specimens and DNA samples to find out.

As I was enjoying my morning coffee on July 2, Poignard emerged from the *perruque* to get my attention. He was agitated and pointing to the east before he found the focus to say something. He surprised me by exclaiming in English, "Boss! Snek!" The rest of the team had not yet emerged from their tents, but the "snek" call got everyone moving. Within seconds, we were staring at a drowned water cobra that a young fisherman had caught in his net near an islet in the lake. It looked very similar to the photo from the German book, but when I looked closely, I thought I could see faint brown bands that had faded into the grayish background color. Overjoyed to have a water cobra from the lake, I rewarded the fisherman and asked him to bring more if any turned up.

He must have bragged to his buddies, because by the afternoon, I had three more water cobras, including one that had been caught alive with assistance from Wandege. I scrutinized the color pattern of each individual carefully, noting how one had more distinct brown bands around the front of its body. Another individual had clearly demarcated brownish bands along its entire body, not very different from the typical form, but it had an odd marbled pattern around its neck. These three-foot snakes also seemed to be small compared with published records of the banded water cobra. Was I really looking at a new species, or were these snakes just smaller color variants of the widespread water cobra?

Unbeknownst to me at the time, a group of respected European herpetologists had already noticed the Mai-Ndombe water cobra in the pet trade, and their genetic analyses of the captive specimens found that they were nearly genetically identical to other populations of the banded water cobra.[75] Soon after I returned to Texas, I sent samples to a colleague in Europe and his preliminary genetic analyses also suggested the Mai-Ndombe cobra was not distinct from the widespread form.[76] Based on these data, none of us was convinced that the Mai-Ndombe cobra was a new species, and the parallel studies were shelved. However, years later, many unanswered questions lingered. Why were the lake snakes so different in size and color pattern? Did the unusual conditions of the lake create strong selective pressures (i.e., ecological factors that favor certain traits more than others)? And how genetically isolated were these snakes from populations of the common banded water cobra?

Answering these questions will require lots of data, sophisticated analyses, and testable hypotheses. Perhaps the lake snakes became isolated during ancient fluctuations in climate that altered connections between the lake and nearby rivers where the widespread form occurred. If the isolated lake population started to change genetically and morphologically, a process that can occur more rapidly when populations are smaller (i.e., genetic drift), they could have started down their own evolutionary path. But then if the climate changed again and the relatively brief isolation ended, they could have interbred freely with the widespread form, possibly explaining the close genetic affinity, at least with the data analyzed so far.

During the lockdown in April 2020, a respected French colleague and his collaborator decided to publish a paper that described the Mai-Ndombe water cobra as a new species, which they dubbed the dwarf water cobra (*Naja nana*). Their study had no DNA evidence, but it included morphological data from many natural history specimens, including other closely related species of water cobras (*Naja annulata* and *N. christyi*). They showed that the dwarf water cobra was less than half the size of the other species of water cobras. They also found some differences in the range of scale counts (e.g., number of belly scales) among the three species. The specimen they chose as the holotype (the representative individual for a new species description) was a handsome male with a black background color, and white and cream blotches and spots. Although they looked at scores of additional specimens, they said that the color patterns were similar to the distinctive male (except for abundance of small white or yellowish dorsal spots), and there was no mention of the bands that I had seen in my specimens. Interestingly, they spoke to a fisherman who said he had been bitten on the hand fifteen times, but in five of these cases, he suffered only minor pain and swelling for one or two days.[77]

Before the widespread use of DNA sequence data in the 1990s, size, scale counts, and color pattern were the main morphological characters that were used to distinguish different species of snakes from one another. But now that we know the dwarf water cobra is nearly genetically identical to the banded water cobra, should we continue to recognize them as distinct species? My Czech colleague Václav Gvoždík recently analyzed an impressive mitochondrial and phylogenomic dataset from this group of snakes, and once again, he confirmed that the genetic data do not support two distinct species. And yet at least some of the differences noted in the description of the dwarf water cobra (especially size) suggest the Mai-Ndombe population could be considered an "evolutionary significant unit" that deserves some level of recognition.[78] One possible solution might be to recognize it as a subspecies, a controversial taxonomic subcategory rarely used by American herpetologists in the modern era,[79] but my colleagues and I will make the final decision when the genetic data are published.

For now, I remain intrigued by the mysterious processes that have shaped the evolutionary history of these snakes. With time and a lot more data, I am confident that future herpetologists will have the tools to unlock these secrets. The only question is, will wild populations of the snakes still exist? Increased heat from climate change has already eliminated Lake Faguibine near Timbuktu, Mali, in West Africa.[80] On the other hand, increased rainfall from climate change has caused Lake Turkana in East Africa to increase in area by 10 percent over the last decade, which caused an ecological catastrophe for people who have lived near it for millennia.[81] As temperature and rainfall patterns around Lake Mai-Ndombe continue to shift in the coming years, it is not easy to predict how the lake and its animals will be affected. But one thing is certain—change is coming.

MAMBA FOR DESSER

If one searches for the definition
of *pygmy* in Google, the primary noun provided by Oxford Languages is,
"a member of certain peoples of very short stature in equatorial Africa and
parts of Southeast Asia."[1] Many hundreds of years ago, Homer's *Iliad* first
mentioned that the Greeks used this term to describe a mythical population of
cubit-high (length of a forearm) people from Africa.[2] Parts of the myth were
proved to be reality, at least in some aspects, once European explorers started
publishing accounts of their adventures in remote areas of Central Africa in
the late nineteenth century.[3] In his best-selling book, *In Darkest Africa*, Henry
Stanley's popular tales of warlike pygmies with poisoned arrows likely colored
the perceptions of generations of Western readers.[4] Ever since then, *pygmy* has
generally referred to Central Africa's short-statured (males average about 4.5 to
5.25 feet), Indigenous hunter-gatherers.[5]

In the Lake Tumba area of western Congo where I worked in 2013, the
average Twa (pygmy) man is 5 feet 3.4 inches, among the tallest of the Indigenous
forest peoples in Central Africa.[6] Several hypotheses related to environmental
factors have been proposed to explain this unusually short stature. Thermoregulation (maintenance of internal body temperature) is difficult in the warm and wet
conditions of the forest, but a smaller body lowers energy requirements. A shorter
stature could also be an advantage for moving through the dense vegetation of
the forest while hunting or gathering food, again curtailing energy expenditures.[7]
These ideas have been challenged by anthropologists and ethnographers, and

further study is needed to reach a firm conclusion. More recent genetic studies suggest a link between short stature and improved immunological function, perhaps as a way to combat heavy exposure to parasites in rainforest environments.[8]

Because of their astonishing ability to live as nomads (or semi-nomads) in beehive-shaped huts in the heart of the forest, a feat assumed to have ancient origins, some historians have hinted that these "primitive" African societies remained unchanged since the late Stone Age.[9] This perception has even prompted some to question whether they are human.[10] In an especially shameful spectacle of American history in 1906, a Congolese Mbuti named Ota Benga was displayed with an orangutan in the monkey house at the Bronx Zoo—tens of thousands of New Yorkers flocked to the zoo to jeer at him.[11]

Another common theme in the historical record is that the forest hunters are subservient to (or even property of) their taller farming neighbors (i.e., Bantu villagers[12]), who often trade agricultural products or money for the hunters' meat.[13] This imbalanced relationship probably originated as the two groups first came into contact during the "Bantu expansion," when people skilled in metallurgy spread late Iron Age culture from grasslands in modern-day Cameroon and Nigeria to most of Central Africa during a dry spell three to five thousand years ago.[14] The asymmetries likely became worse in the late nineteenth century, when Bantu used violence and intimidation against their hunter-gatherer neighbors to obtain forest resources for white colonists.[15]

But some of these assertions are oversimplifications or fallacies that have been warped by the lens of colonialism, racism, and outdated assumptions.[16] For example, many groups of Bantu farmers will hunt wild animals, and some forest people have shifted at least some of their way of life to agriculture, blurring some of the lines between these groups.[17] Sometimes these seemingly disparate groups live together in intermingled, multiethnic communities.[18] Many groups of forest people are indeed marginalized or discriminated against by their Bantu neighbors, and yet they are perceived to be healers and religious experts with supernatural powers, including invisibility and the ability to turn into an elephant![19]

Given this history, it should not be surprising that some consider "pygmy" to be pejorative, and the Republic of the Congo (neighboring Democratic

Republic of the Congo) has tried to outlaw the term. Academics and the people themselves have not reached a consensus on an appropriate replacement name, and for now, the easily recognized moniker continues to be in widespread use in Africa and around the world.[20] Reporters working in western Congo noted that some individuals continue to introduce themselves as *pygmies*.[21] In my experience in Congo, the term is used without compunction by a diverse array of ethnic groups, and I have rarely witnessed its use in a derogatory manner. Nevertheless, I hope an appropriate substitute can be found in the future, preferably chosen by the African peoples involved, but this will be no easy task. The Congo Basin includes at least fifteen different ethnolinguistic groups[22] with over a quarter of a million of Central Africa's forest peoples,[23] making them the largest and most diverse group of active hunter-gatherers remaining in the world.[24] Below I refer to the *Batwa*,[25] the specific name for forest people in much of western Congo.[26] Thanks to Colin Turnbull's moving book that emphasized their egalitarian societies, extraordinary talent with music and dance, and unrequited love for their forest home, Central African hunter-gatherers are also known more broadly as "the forest people."[27]

Understanding the true origins of forest peoples is hampered by a lack of paleoarchaeological data in forested areas of Central Africa (the acidic soil stymies preservation of bones), which one prominent scholar likened to a historical terra incognita.[28] However, inferences from genetic data and cultural anthropologists can provide clues to the ancient past. Genetic studies suggest an ancestral group of forest people formed their own branch of the human family tree roughly 70,000 years ago, except for those in Cameroon and Gabon, who diverged only 3,000 years ago, perhaps as a result of the Bantu expansion. Intriguingly, a fair amount of genetic differentiation exists between different groups of forest peoples. One study noted a divergence between western and eastern populations of forest people happened about 20,000 years ago during the Last Glacial Maximum, when relatively cold temperatures caused rainforests to contract into smaller refuges. If the forest peoples stayed within their isolated forest refuges during this time, it could explain how they ended up with distinctive genetic signatures.[29] Despite widespread taboos regarding intermarriage between forest peoples and Bantu tribes, all of the Central African populations showed evidence of at least some admixture, or interbreeding, between these populations that had been previously isolated, but much of this happened in the last 1,000 years.[30] A significant caveat is that these population genetics studies have huge gaps in sampling, especially in Congo, and our understanding of these trends might shift with improved analyses in the future.[31]

An interesting window into the beliefs of the Batwa and their way of life in western Congo was provided by letters from an American missionary doctor named Jerry Galloway, who lived among them from 1980 until his death in 2007. Working in a remote area northeast of Lake Mai-Ndombe, Galloway observed how the Batwa blamed sickness and death (including AIDS from 1987 onward) on everything from curses by enemies to aggrieved ancestors and spirits to sorcery. If one of them disappeared in the forest, they firmly believed that a cannibalistic Bantu boogeyman called the Bengondu killed and ate the victim. He also witnessed how the Bantu tribes mistreated the Batwa, sometimes treating them as slaves. When Batwa women encountered Bantu men on forest trails, the women would enter the forest, turn their backs to the men, and wait patiently for them to pass. As for their diet, the Batwa "ate everything that moved," including crocodiles, turtles, and, as I would later confirm, snakes.[32]

Nursing a hangover from a palm wine sendoff party the night before, we carried our gear from Mpote-Emange to Inongo, consolidated it into a pile at the edge of a boisterous market, and watched Chifundera disappear on his mission to find a fisherman who had agreed to ferry us to the northeastern shore of the lake—for hundreds of dollars. Dozens of vendors sold a variety of sundried fish on open tarps, and there were also smoked hunks of meat from terrestrial animals, crates of soda and beer, clothing, plastic wash basins, and other cheap items that seemed to originate from Asia. Knowing this might be the last market we would encounter for several days, if not weeks, everyone left me on a shadeless hill to stand guard while they shopped for last-minute luxuries like soap, snacks, and alcohol. Squinting in the mid-morning sun and already sticky from the heat,

I marveled at how people managed to stay upright as they shimmied through slippery mud on uneven ground, all while wearing cheap sandals with no traction.

By the time Chifundera returned and our gear was loaded into the fisherman's boat, my shirt was soaked in sweat, but I was excited to be on the move again. A stiff breeze from the lake cooled our bodies as we headed north. Chif said we would head to a small village called Isongo (a.k.a. Isanga,[33] not to be confused with Inongo) on the northeastern shore of the lake, where the fisherman said a road led north to Lake Tumba and Mbandaka, the largest city in northwestern Congo.

As our boat approached Isongo, curious residents congregated at the shore to witness the rare visit of a white man to their village. The captain told us they were from the Bolia tribe, and they wore typical garb of the average Congolese citizen—old plastic flip-flop sandals, T-shirts, and faded pants or cutoff shorts. Many of the shoeless children had pot bellies, accentuated by open shirts that had lost all their buttons, and some of them wore nothing but dingy underwear. The children whooped and hollered "*Mondele!*" (White man!) in excitement as we landed on the shore.

Somehow during the raucous welcome, my eye was drawn to the edge of the crowd, where a smaller group of teenage boys observed us with reserved curiosity. But there was something about them that seemed different. They were remarkably similar in height, with large and bright eyes, wide noses, thin lips, and wiry bodies. They were expressionless, but somehow I could sense their humble confidence. Then it hit me. They were Batwa. Aristote had followed my glance, saw the boys, and exclaimed, "Pygmies!" I am sure they heard him, but they made no reaction.

As Aristote hopped off the boat, something at the edge of the water caught his eye. Following his glance, Wandege and Chifundera dropped into the stooped posture of hunters, and the three of them slowly surrounded a clump of reeds. Too surrounded by the mountain of baggage in the boat to help them in time, I leaned over the edge and watched. Puzzled, everyone on the shore froze to stare at us, whispering wild theories about our purpose. With a kung-fu-like swipe, Aristote grabbed a rocket frog (*Ptychadena*), and held it up triumphantly. Wandege and Chifundera smiled, I clapped, and the crowd roared with approval. When I glanced back to the Batwa, they had barely moved, but one of them was smiling.

The Bolia peppered us with a million questions as we transferred our gear to the center of the village, where they allowed us to set up our camp in a grassy clearing. Knowing it would be unwise to wander around at night before the chief of the village and other locals had been informed of our purpose,[34] Chifundera visited the village leaders while we chatted with the large group of people who had come from near and far to stare at us. After being the center of attention and butt of many jokes for an hour, I retreated to my tent and drifted off to sleep.

I emerged from my tent after the morning sun had warmed it to an uncomfortable temperature to find Wandege taking photos of Aristote as he posed with a colorful flower. "Feeling artistic today, Aristote?" I asked with amusement.

"Nooooo," he replied emphatically, in the way that he always did when I said something to affront him. "This *flor* was a gift from my new pygmy friend, and it is very special."

"Really?" I said with surprise.

"Yes, of course," he continued. "Yesterday there was a ceremony to celebrate the birth of one children. The pygmies were singing and dancing." The festivities must have happened some distance from the village, because I had not heard a thing.

"*Pombe*?" I asked.

He glanced at Wandege, who smiled back with a guilty expression.

"Yes, of course!" Aristote replied with a grin. "I took some *pombe* with one pygmy man. The man said ha! As I give you this *flor*, my God will bless you, your job will be ok."

"Oh, that is really very nice," I said while realizing the significance of the gift. I looked more carefully at the flower and saw that it had an unusually long, brown stem, a reddish-orange base, a corklike center, and a bulbous orange-yellow top with hot pink highlights. It resembled a beautiful microphone, and I shared my thought with Aristote.

"Yes, of course!" he exclaimed. "The pygmy told me that it is a magic microphone. When you use this flor to ask for something, God must hear it and say yes."

Based on Wandege's photo of it, the microphone would eventually be identified as the inflorescence (the complete flower head including stems) of *Parkia bicolor*, the African locust-bean tree.[35] A member of the legume family, this tree reaches heights as tall as 130 feet, and it has a large range from West Africa to eastern Congo, mostly in lowland rainforest.[36] The tree is pollinated by bats, and because it has edible fruits and seeds, it is often left standing when other rainforest trees are felled to clear land for agriculture.[37] According to the Plant Resources of Tropical Africa website, the tree's bark, leaves, and roots are used in traditional medicine to treat everything from toothaches to sexually transmitted infections.[38] In the hilly forested areas of West Africa, "sabre-toothed frogs" of the genus *Odontobatrachus* seem to purposefully ingest pinnate leaflets from the trees, a rare occurrence among frogs, but the leaves do not seem to provide nutrition and the reason for their ingestion remains a mystery.[39]

My mind drifted to thoughts of what I might say to the man upstairs if the microphone really worked, but Aristote quickly refocused my attention.

"I have become friends with the pygmies, and they say they know a stream where many snakes live in the forest. They can take us there today," he said.

I knew from previous experience in eastern Congo and Burundi that the forest people can be extremely helpful in finding rare species, and I immediately agreed to accept their offer of help. Too excited to wait for us to plan our foray into the forest, some of them emerged from it just minutes later with colorful snakes I had never seen before. Eight-inch ones with dark brown bands on a bright orange or yellow background proved to be juvenile ornate African water snakes (*Grayia ornata*). They flickered tongues that had a black base and flaming orange forked tip. Several specimens of another foot-long species had unusually small, beady eyes, nostrils oriented toward the top of the snout, with orange and brown bands that became less contrasting along the body until they resembled a checkerboard pattern on the tail. Lacking a detailed library to help me identify them in the jungle, it would take a year for me to determine that they were Schouteden's sun snake (*Helophis schoutedeni*), a rare species that had been known from only three published specimens since 1937. Now we know that this species is relatively widespread in aquatic habitats of the lowland rainforest in Congo, but virtually nothing is known about its ecology, behavior, or evolutionary relationships.[40] This was a fantastic start, but the best was yet to come.

Our guides helped us find more of the interesting frogs we had seen at Mpote-Emange, including a possible new species of toad (*Sclerophrys*), clawed (*Xenopus*) and dwarf clawed frogs (*Hymenochirus*), white-lipped frogs (*Hylarana*), and a unique rocket frog (*Ptychadena*) with rust-colored legs that was never encountered again. Soon we had so much material that I had to set up my laboratory in a large mosquito shelter so that we could work while shielded from the abundant insect life near the lake. Wandege and I were engaged in an extended photo shoot of the frogs when a sweat-drenched Batwa girl approached us with an old tin can. She handed it to Wandege, who raised his eyebrows in a way to tacitly imply, "Well, that's something you don't see every day."

In his quiet, understated manner, Wandege offered the can to me and asked, "Can you see one *Polemon*?"

I gasped in recognition of the snake genus name. *Polemon* belongs to a group known as collared snakes (subfamily Aparallactinae), many of which have a light-cream, yellow, or orange ring around the neck. These snakes have a large amount of variation in their fang position and structure, if they have them at all, depending on the type of prey they consume. One rear-fanged group (*Aparallactus*) favors centipedes and earthworms. *Polemon* eat other snakes, sometimes close in size to their own body,[41] and their large and deeply grooved fang is located toward the front of the mouth where it is used to envenomate prey.[42] Because *Polemon* are secretive burrowing snakes, they are encountered relatively rarely, and snakebites to humans are virtually unknown. However, one especially colorful, striped species from West Africa (*Polemon acanthias*) caused moderate swelling and blister formation after biting a herpetologist on the finger.[43]

When I looked in the Batwa's can, I saw a black, ten-inch-long snake with a creamy yellow collar that culminated in a jagged pattern resembling flames on the crown of the head. I had never seen another snake like it, and for a moment, I was so mesmerized that I did not notice the large crowd of Bolia who were surrounding us to stare at the serpent. Years later, after lots of detective work, the snake would be identified as *Polemon robustus*, which was described during World War II from specimens found near Lake Mai-Ndombe.[44]

I thanked the girl for bringing us such a rare treasure, but the boisterous chatter from the Bolia suggested they did not like the attention she was receiving from us. Wandege glanced at the crowd with concern as he listened to their animated comments. When I looked at them, I saw people gesturing in ways that seemed to be venting frustration. Poignard eyed them with a hardened look, almost as if he was mentally preparing for a fight.

"Wandege, what's wrong?" I asked.

"I sink now zeese peoples can be jealous of da pygmies," he said in broken English.

"But why?" I asked. "They are only trying to help us."

"Maybeee," he continued. "But I don't sink day likes it."

Chifundera and Aristote returned from the forest with their Batwa guides in the late afternoon, and the frog hunting had been good. Both of them were quick to notice the tense mood of the people in the village and the Batwa retreated immediately. It took us only a few minutes to decide that it was time to leave Isongo, but Chifundera would need time to find a truck to transport us to the north. He suggested we move to one of the exclusively Batwa villages near a town called Bikoro where he had worked a few years before. He said the Batwa were friendly and helpful and we would not have to deal with other tribes that might resent our interactions with the forest people.

By dawn, word had spread that we planned to leave, and the mood was subdued as we packed up to depart. Several Bolia had told Chifundera that it was impossible to find transportation, but undaunted, he left on a motorcycle to search for a truck. Hours later I was sitting on the side of the road leading into Isongo when Poignard appeared at the front of a procession of teenage

Bolia boys. Their upright posture and confident swagger suggested they had important news.

"Boss, crocodile," said Poignard with a French accent as he gestured with a flourish to a group of boys that were marching out of a nearby patch of forest.

Three of the strongest boys approached with four-foot-long dwarf crocodiles (*Osteolaemus osborni*)[45] that had been hogtied in a way that their jaws were fastened shut and attached to the tip of the tail, forming a convenient, ringlike shape that could be worn like a cruel purse over the shoulder. They surely thought that I would pay a king's ransom for the reptiles, and their emergence from the forest seemed staged to make me think they had just returned from a hunt. But I knew better—the Batwa had certainly found and captured the crocodiles from deep in the forest and probably sold them to the Bolia for meat. Poignard rubbed his hands together in anticipation of a feast. Everyone stared at me in disbelief when I sighed, shook my head, and showed no interest whatsoever. Rebuffed, the boys snatched up their crocodilian prizes and headed for the river, where the hapless animals would be sold to a middleman and begin their torturous journey to dinner plates in Kinshasa.

Dwarf crocodiles dig burrows just above the waterline in swamps and slow-moving bodies of freshwater in rainforests, where they emerge at night to hunt for a diverse array of prey, including fish, amphibians, and crustaceans. When they can catch them on trails or ambush them at the water's edge, they will also eat snakes, birds, and mammals.[46] Like some other crocodilians (alligators, caiman, crocodiles, and gharials) throughout the world, African dwarf crocodiles are considered to be at least "vulnerable" to extinction, but the most recent conservation assessment from 1996 likely underestimates the extinction risk currently faced by the species in Congo.[47] Because they are relatively small and less dangerous compared with other African crocodiles, dwarf crocodiles are hunted for their meat wherever they occur. A Belgian woman who ate crocodile in Congo described the meat as a "delicate flavor, like veal."[48] Crocodiles are resilient animals, and they can be transported long distances to markets, even when exposed to the sun without food or water. Tens of thousands of them are illegally poached and sold in urban "bushmeat" markets in Central Africa every year, and this is likely destabilizing entire ecosystems.[49]

Hours later Chifundera finally showed up in a wheezing truck that backfired in a cloud of dust as it rolled to a stop near us. I was about to ask a rhetorical question about the safety of our transportation when Chifundera whipped open a squeaky door and barked "*Twende!*" (Let's go!). Knowing we had no other choice, we loaded the back of the truck with our gear and waited patiently while the middle-aged driver tried in vain to start it. After multiple failed attempts, he commanded some of the remaining Bolia boys to push the truck to help him kickstart it. Happy to have something to do, the boys helped us rock the vehicle back and forth until we managed to gain enough momentum to push it forward. With a shrieking clank, the truck sputtered back to life, I thanked the boys with enough money to buy some sodas, and they waved goodbye as we left Isongo behind.

We drove north in silence for hours as we passed small remnants of forest on the rough road. A sweaty hour was spent extracting the truck from a muddy ditch and kickstarting it again after it slipped in on a sharp turn. Hot and parched from the late afternoon heat, we entered Bikoro on the eastern shore of Lake Tumba to find a small town with many old Belgian houses and buildings. In the colonial era, Bikoro was known as a "holiday and rest resort" because people swam in the lake, the Hotel du Lac boasted famous cuisine, and the town had a primary school and hospital.[50] More interesting to me was the scientific research station that had been established just after World War II by the Belgians at Mabali, a few miles to the south.[51]

Less than a decade before my visit to Bikoro, at the tender age of seventy-three, a Danish herpetologist and world authority on African reed frogs (*Hyperolius*), Arne Schiøtz, had spent a few days at Mabali. He was not disappointed by the diversity of reed frogs he encountered near the eastern shore of Lake Tumba, but the most interesting find turned up in an unexpected way. Children brought him an unusually large female reed frog from roadside herbs near the station, which had a unique pattern of chocolate brown with lime-green flecks on its back and a bright orange belly. In a published paper about his discoveries, he seemed convinced that the "strangely coloured *Hyperolius*" represented a new species, but without more specimens for comparison, he remarked it would be unwise to describe it.[52] I fantasized about finding more of the striking frogs, and pondered whether DNA results, unavailable to Schiøtz, would confirm their status as a new species.

My growling stomach refocused my attention on more immediate concerns. After an hour of searching, the only restaurant we could find, Chez Maman Paulette, was one that specialized in bushmeat, including monkey and antelope. Luckily, they also had fresh fish from the lake and everyone ordered it. They also had the rare luxury of electricity from a generator, and everyone but me ordered beers. I quenched my thirst with a grenadine soda that was so cold it seemed to freeze the inside of my body as I drank it, but after weeks of lukewarm water, the sensation was indescribably pleasurable.

Relaxed and contemplative, I scrutinized the remains of an old Belgian house in the distance. Curious, I asked Chifundera if he had interactions with the Belgians during the colonial era of his childhood. He told me about a tall Norwegian man named Langseth, who was working as an inspector of primary and secondary schools in the Department of Education with the Norwegian Free Mission (a Protestant organization) in the Belgian Congo. During the inspection, Chifundera's teacher presented him as an excellent student. After a brief observation, Langseth agreed, exclaiming, "You! You! You must go to high school to learn more!" And so he did, eventually earning a college degree that would land him a coveted position as a herpetologist at the Centre de Recherche en Sciences Naturelles in eastern Congo.

Darkness descended on us as I paid the bill, and our attention turned to the question of a safe place to spend the night. The obvious choice would have been to find a place in town because it is dangerous to travel in Congo at night. But three years previously, Chifundera had worked at a Batwa village just outside Bikoro, and he suggested we could find one near the road. Many groups of forest peoples in Central Africa were encouraged or even forced to relocate their villages to roadside locations in the colonial era and subsequent decades for nefarious reasons, including easy access to labor and tax collection.[53] Unfortunately, locations near roadsides are associated with degraded forest, less suitable habitat for animals that are hunted by the forest people, and other negative effects that can have severe consequences on the only way of life they have ever known.[54] I quickly weighed the pros and cons and decided we would give it a try. After all, I reasoned, if it did not work out, we would be close enough to Bikoro to find another ride as we continued our journey north.

The afterglow from our dinner quickly evaporated when we had to kickstart the truck several times to get it started again, but after traveling only a few minutes out of town, we found a Batwa village called Npenda that looked suitable. A quick shine of my headlamp showed that the vegetation immediately adjacent to the road had been heavily damaged by human activities, including cultivation of banana plants, but not far in the distance, it looked much better.[55] It sounded better, too—I did not have to strain my listening abilities much to hear an impressive cacophony of insect life, suggesting a healthy ecosystem. Between the forest and road, I could see a handful of mud-and-reed shelters with cooking fires that cast an orange glow through cracks in the walls. The arrival of our truck caused a flurry of activity from the alert residents, and several doors flew open, expunging small clouds of smoke, pets, chickens, and people. A surprisingly tall figure emerged from one of the structures, and recognizing that he must be the chief, Chifundera greeted him and started to explain our business.

The entire village seemed to surround the two men by the time the conversation was over, and with a quick thumbs-up gesture, Chifundera indicated that we had permission to set up camp. Grateful to be free of the problematic truck for good, we unloaded our gear and started to pitch our tents near the edge of the forest. Even though we had arrived unannounced after dusk, the Batwa welcomed us with friendly smiles and immediately offered to help us with our tents. The children were incredibly excited to have visitors, and they danced and screamed as our headlamps illuminated them in brief glances during our work.

Once we were settled, Aristote said it was ok for us to take a quick peek in the forest behind the village, and two young men would guide us. Within minutes, we found several unknown species of frogs, including two kinds of green forest treefrogs (*Leptopelis*), white-lipped frogs (*Hylarana*), and a small toad (*Sclerophrys*) with an unusually long paratoid gland—a toxin-secreting structure behind the eye of most species of toads that provides a defense against predators. One of the reed frogs that Schiøtz had found at Mabali, the fantastic reed frog (*Hyperolius phantasticus*), turned up too. Some individuals had light green bodies, yellowish lines at the border of their flanks, grayish-blue legs, and their bellies were translucent, allowing one to see some of their internal organs and circulatory system. Others were orangish brown with yellow spots, and their underside looked like they had been dipped in dark brown paint. Convinced we had chosen a good spot, we returned to the village and showed the frogs to the children, who shouted and reacted with excitement while their parents laughed. The people seemed happy and curious, and I fell asleep with a feeling that I had made the right decision to come to Npenda.

Shortly after dawn, I emerged from my tent to see that a crowd of three dozen women and children had already gathered near the road to watch me. Many of the children were barefoot and clad in dirty clothes that had so many holes they barely passed for rags. A handful of women in their twenties watched from the rear, and two of them held babies who stared with their mothers. One teenage girl had an enormous woven basket strapped to her back, probably to gather firewood. Their expressions ranged from curious amusement to apprehension. When I smiled and waved, their eyes grew wide and their jaws dropped, and as I turned toward the forest to answer nature's call, I could hear their animated conversation in my wake. Aristote greeted me on my return visit from the village latrine, an open pit with a few withered palm fronds surrounding it. I remarked in passing that the vegetation did not offer me much privacy because I was taller and bigger than the average person in Npenda, and I had tried to ignore the laughter as the crowd watched me. I did not think anything of it and was prepared to sacrifice my modesty, but within an hour of Aristote passing along my sentiments, the Batwa constructed a new shelter around the pit that was large enough to accommodate my body. Embarrassed but thankful, I passed out candy to the children and furnished some gifts to the village elders. Everyone seemed to be delighted.

By the time Wandege and I had erected the mosquito shelter for my laboratory, most of the village had gathered to watch, and judging by the presence of a few teenagers on bicycles, word of my presence had spread to nearby villages too. The crowd swelled when I took out my camera to photograph the frogs we had found the previous evening, and they were shocked by the bursts of light emanating from my flash. At the edge of the crowd, I caught a glimpse of Poignard, who was manhandling some of the rowdiest people in a vain attempt to control them. For a moment I was so shocked that I did not react, but then I felt outrage because of his shameless behavior of putting his hands on people in their own village.

"Hey, Poignard, stop!" I yelled as I pointed my finger at him. *Stop* was a word that needed no translation, and as everyone turned to look at him, he released his grip from a teenage boy and grimaced nervously. Chifundera patted his shoulder reassuringly and said something to him in Mashi. Realizing that the crowd had become untenable near my workspace, Chifundera produced his own camera, took pictures of the people, and dazzled them with the playback option to show them their own photos. Screams of delight emanated from young and old. Poignard sulked on the sidelines, chastened by my reprimand. His expression seemed aggrieved as if I had overstepped my authority by disrupting his perceived right of dominance over the Batwa.

I was still contemplating the injustice of Poignard's mentality when Aristote appeared with three Batwa guides, all of them covered in sweat. "Jambo (hello), boss," he started. "We found lizards basking in the morning sun. Can you see them?" He held out containers with wriggling lizards that flashed brief glimpses of vibrant turquoise.

Distracted and intrigued, I immediately forgot about Poignard and accepted the containers for a better look. They contained a handful of frogs and dull brown day geckos (*Lygodactylus*), which are quick and difficult to catch. But then I found the container with the turquoise lizards and gasped in disbelief. Somehow the Batwa had managed to capture three western blue-tailed gliding lizards (*Holaspis guentheri*). About five inches long from head to tail, the lizards had striking black and yellow stripes on the head that transitioned to a series of black and blue stripes on the back. The bluish color became increasingly vibrant toward the back of the animal until it resembled a neon turquoise hue on the tail, flanked by canary yellow scales at the fringes. Thin cream stripes decorated the black flanks. The lighter areas of the lizard's pattern are known to have ultraviolet fluorescence, which is visible to other lizards and possible predators like birds.[56]

As their common name implies, the lizards are capable of gliding nearly one hundred feet from tree to tree, at times curving their back to steer the "flight" to a specific target.[57] They can do this because of an extraordinary example of adaptive evolution—the lizards have a flattened body, triangular scales on the lateral edge of the tail, a relatively small body mass, and low-wing loading from modifications to their skeleton—all of which allow them to slow their descent after a jump or fall, stay aloft longer, and reduce impact forces as they land.[58] Although they are known to venture to the ground, gliding lizards seem to spend most of their time hunting for insects and spiders in the bark of trees during the day, sometimes venturing to heights approaching one hundred feet.[59] The Batwa must have been extraordinarily lucky and stealthy to find them within arm's reach of the forest floor.

It was impossible to hide my excitement about the extraordinary lizards, and the Batwa seemed to instantly recognize my passion. I beamed at the men who had found the lizards, and they smiled back in tacit recognition of their intimate knowledge of the forest and its inhabitants. Chifundera, Wandege, and Aristote took turns helping me in the lab while the others ventured into the forest with the Batwa to look for more species.

Even after the novelty of my presence at Npenda had faded away, three curious boys, Mokanga, Bonkumu, and Bakota, came to check on me frequently. Wearing dirty shorts and nothing else, they watched me work, swatted flies away from their faces, and whispered to one another with amused, furtive glances. I guessed they were about twelve, but the tallest of them barely reached Aristote's chest. He struck up a conversation with them and asked what gift they planned to bring for me.

"*Pili pili!*" (Hot chiles!) suggested Bakota with a smile, to which his companions responded with an affirmative "Mmmm," as if to say "Good idea!"

I gave them some candy and Bakota showed off his gymnastics skills, including a series of running front flips that would have impressed Simone Biles.

During the normal hunting activities of the Batwa, we had opportunities to see other rare animals. Aristote brought my attention to two Batwa men who had captured one of the Congo's most extraordinary creatures. They carefully plucked a long (1.5 feet), scaly tail from a bucket, and at the end of it, curled into a ball like a roly-poly bug, was the grayish, armored, foot-long body of a tree pangolin (also known as the white-bellied pangolin, *Phataginus tricuspis*). Slowly, the animal allowed itself to unfurl like the frond of a fern until I could see four limbs with long, recurved claws, small eyes rimmed with white furry eyelids, and a long, pointy snout that ended in a doglike nose. Except for its face, the entire body was covered in large, striated scales that reminded me of artichoke leaves. It did not make a sound and was perfectly still, except for its nose, which twitched rapidly as it smelled the air.

African tree pangolins are one of eight pangolin species that occur in Africa and Asia. Weighing about two to three pounds, the arboreal tree pangolins are much smaller and lighter than the closely related giant pangolin (*Smutsia gigantea*), a terrestrial species that can exceed six feet and weigh over seventy-five pounds. Tree pangolins have tongues that are as long as their bodies, and they use them to slurp up termites and ants, including formidable army ants. To avoid painful attacks by soldier ants, tree pangolins will move quickly, "shiver" their scales, and use their long, prehensile tail in a sweeping motion to concentrate the ants for efficient feeding. Most of this feeding takes place at night on the ground, and during the day they shelter in tree hollows that can be up to fifty feet off the ground.[60]

Like many other African mammal species, tree pangolins are hunted for their meat, and they are one of the most common species in bushmeat markets.[61] But they are also utilized in African traditional medicine, including putative treatment of a wide array of diseases related to asthma, cardiovascular and dermatological problems, infertility, rheumatism, and back pain. In some parts of Africa, pangolin parts are even used to confer invisibility, concoct love potions, and deter witches and evil spirits.[62] Unfortunately, additional bogus beliefs about pangolin scales are popular in traditional Asian medicine in a similar vein to rhino horn, and despite all eight species being listed as Appendix I (the highest level of protection) under the Convention on International Trade of Endangered Species of Wild Fauna and Flora (CITES), pangolins are now one of the world's most trafficked terrestrial vertebrates.[63] A recent genetic study of illegal pangolin seizures in China confirmed that tree pangolins are the most heavily poached pangolin species (88 percent) in Africa.[64] The scale of the problem is enormous—over a roughly ten-year period in China, seizures of illegal African pangolins and their products (i.e., meat, scales) represented over 143,000 animals, and this number is likely the tip of the iceberg.[65]

Later that day, a breathless Aristote showed up at the lab waving an enormous snake tail at me to get my attention. It was lime green with black-edged scales in a distinctive grid-like pattern, and given its size, I immediately knew it belonged to the deadly Jameson's mamba (*Dendroaspis jamesoni*). Because these long snakes (up to 8.5 feet or more) spend most of their time in trees where they hunt arboreal rodents, lizards, and birds, they rarely come into contact with people. Few documented bites are known, but in some areas of Africa (including Congo) mambas are hunted for meat and unreported casualties are likely. In Nigeria where palm oil tappers must climb up to sixty-five feet to harvest fruit, some have fallen to their deaths when they encountered Jameson's mambas and other arboreal snakes, which are quite common in the plantations.[66]

Mambas are in the Elapidae family, the same as cobras and American coral snakes, and the neurotoxic venom of this group of snakes is arguably the deadliest of all. Mambas have two types of toxins that are unique to them, dendrotoxins and fasciculins, which affect neuromuscular junctions (connections between nerves and muscles) and lead to paralysis and tremors.[67] Other symptoms can include swelling of the entire limb, vomiting, numbness, blurred vision, slurring of speech, and cold and sweaty skin.[68] Somebody took a huge risk to capture that snake, and I felt a surge of adrenaline as I examined the tail. Based on its size, the animal must have been at least six feet long.

"Aristote, how did you get this?" I asked with an incredulous expression.

His eyes grew wide as he told me the story. Somebody had seen the snake moving at the edge of Npenda and had dared to kill it with a machete. Because mambas are among the world's fastest snakes, this one managed to escape, minus the tail. But this initial account turned out to be false, or at least incomplete. Meanwhile, news of the kill spread rapidly, and much of the village showed up for a mamba barbeque. It had taken all of Chifundera's charisma to persuade them to let us have the tail so that we could get a DNA sample and associated voucher specimen for research.

A few minutes later a man arrived with the snake's head, and he conveyed the full story. Apparently, a married couple had been cultivating a field at the edge of the forest near Npenda when the woman saw the snake. Ignoring the danger, both of them used their farming tools to kill the snake, because its meat is considered a delicacy, and the head is used in traditional medicine. I now regret that I did not take the time to conduct a more detailed interview about the potential medicinal properties of the snake. Rightly or wrongly, the scientific community values peer-reviewed publications substantially more than oral stories of Indigenous peoples,[69] but they have centuries of knowledge that we do not and their insights might be a crucial first step to formulate hypotheses for future scientific investigations and improve conservation outcomes.[70]

Many of us view venomous snakes as dangerous animals that should be avoided, and this visceral ophidiophobia is likely an ancient, instinctual fear that guided the evolution of primates, including us. The so-called Snake Detection Theory (SDT) posits that Old World monkeys and apes (i.e., Catarrhini) coevolved with snakes as predators of our ancient ancestors (i.e., constricting snakes like pythons) and later as deadly threats (i.e., via defensive strikes of venomous snakes) in Africa and Asia for so long that our mammalian brains evolved key adaptations for "visual system expansion" to quickly identify and avoid snakes. In evolutionary biology, natural selection (essentially things that can kill) is a powerful force for change, and proponents of the SDT argue that snakes gave us quintessentially human qualities over time, including orbital convergence (i.e., binocular vision), improved visual specialization (for color and object recognition), and brain expansion, especially the areas that control learning associated with fearful stimuli like snakes. A recent study documented increased resistance of Afro-Asian primates, especially African apes and humans, to α-neurotoxins (a powerful venom component) of terrestrial and diurnal cobras, supporting the concept of a coevolutionary arms race between these mammals and reptiles, and at least some aspects of the SDT.[71]

Given that we have been evolving for tens of millions of years to avoid snake envenomation, it is ironic that research into the manifold bioactive cocktails of peptides in snake venoms has led to several life-saving drugs. The angiotensin-converting enzyme (ACE) inhibitor drug captopril, used to treat high blood pressure, was the first FDA-approved drug developed from animal venom in 1981, which in this case was the South American Jararaca Viper (*Bothrops jararaca*). Captopril and its related family of ACE-inhibitor drugs have saved millions of lives.[72] Acute coronary syndrome (ACS) is associated with a suite of conditions that can lead to a sudden drop in blood supply to the heart, including a heart attack, and every year about seven million people are diagnosed with ACS worldwide, including one million hospitalized people in the United States.[73] Tirobifan is an anti-platelet drug used to treat ACS, and it is derived from a toxin of the Asian saw-scaled viper (*Echis carinatus*).[74] Even venomous lizards have made game-changing contributions—the gila monster (*Heloderma suspectum*) from Arizona, New Mexico, and Sonora, Mexico, contained a peptide that led to exenatide, a revolutionary drug that stimulates insulin production from the pancreas when diabetics (10 percent of the US population) experience a spike in blood sugar levels.[75] After pharmaceutical companies noticed that patients taking exenatide and similar drugs experienced the "side effect" of considerable weight loss, further research eventually led to the development of the wildly popular diabetes/weight-loss drugs Ozempic, Wegovy, Mounjaro, and Zepbound.[76]

The list of venom-derived drugs goes on and on, but many more promising ones are in the pipeline of preclinical and clinical trials. Crotamine, a toxin isolated from the South American Cascabel rattlesnake (*Crotalus durissus*), targets rapidly proliferating cancer cells while sparing normal ones, and it also has promise as a painkiller and anti-inflammatory drug. Cenderitide, a peptide derived from venom of the eastern green mamba (*Dendroaspis angusticeps*), is in clinical trials to treat heart failure. Mambalgins, peptides found in the venom of the extremely deadly black mamba (*Dendroaspis polylepis*), are being used to develop powerful, nontoxic analgesic drugs.[77]

Could some as-yet unstudied component of Jameson's mamba venom contain the next breakthrough drug discovery? What about the other poorly studied and potentially new venomous snakes I encountered, including the collared snake (*Polemon*) and burrowing asp (*Atractaspis*)? Absolutely, but scientists have their work cut out for them. Venom can be surprisingly complex, with some snake species containing hundreds of bioactive compounds, and there are thousands of known protein isoforms (variants of the same venom gene) with varying biological activity. Even more variability in venom toxins can occur within a snake species because of factors related to its age, sex, population genetics, geographic location, and type of prey.[78] This variation also has important consequences for the treatment of snakebite, a "neglected tropical disease" designated by the WHO that leads to approximately 1.8 million annual snakebite victims with 138,000 deaths worldwide.[79]

But before bioprospecting for new drugs can happen, the existence of a given venomous snake species, including its updated taxonomic status and name, must be known by the scientific community. Natural history collections from remote, poorly explored areas like Congo are absolutely essential to this process of discovery, underscoring the continued need for expeditions like mine. I would not have to wait long for the jungles of Npenda to yield another truly spectacular serpent.

Over the next three days, the people of Npenda continued to help us find an incredible diversity of amphibians and reptiles, and the snake hunting was especially fruitful. I knew Wandege had found something truly awesome—and venomous—when he emerged from the forest with a spring in his step. His cloth snake bag was tied to the end of a long branch to ensure it was kept at a safe distance from his body. Three excited Batwa followed in his wake and all of them seemed eager to see my reaction.

"Venomous?" I asked as he approached.

"Yesss," he replied with a satisfied grin.

"What is it?" I asked impatiently, like a kid who could not wait to unwrap a present.

Wandege smiled as he carefully untied the bag and then used a pair of snake tongs to pour three feet of wriggling serpent coils onto a muddy clearing near the laboratory. The snake was silvery grayish brown with brown speckles and a series of darker brown diamonds along the entire length of its body and tail. The center of each diamond was light pinkish cream. In an instant, it righted its body, raised its head to reveal catlike eyes, cocked its head back in a threatening, S-shaped coil, and flickered an orangish-red tongue at us.

"Wow!" I exclaimed as my camera documented the snake in all its splendor. It was a powdered tree snake (*Toxicodryas pulverulenta*), so named for the pattern of spots and speckles on its body. Known from forests and nearby habitats across a huge part of tropical Africa, this nocturnal snake eats frogs, lizards, and rodents when it can catch them in trees. These snakes have fangs at the back of their mouth that are capable of injecting venom, but the effect on humans is completely unknown, and given the secretive nature of the species, no snakebites have ever been documented by scientists. Wandege was wise to be cautious because a close relative of the species has a powerful neurotoxic venom that can kill a mouse in a minute flat. None of us wished to be a guinea pig for the first known snakebite. Years later, DNA results from this snake would show that it is an example of a cryptic species (genetically distinct from its superficially similar sister species on the western side of the Niger River Delta), and in 2021, we described it as a new and distinct species, *Toxicodryas adamantea*.[80]

My mind was still swirling with elation when Aristote approached me with Mokanga, Bonkumu, and Bakota. Their eyes were twitching with excitement, and they gestured toward the road.

"Eli, grab your video camera, quick!" he said.

"What happened?" I asked during my mad dash to my tent, ready for anything.

Aristote and the boys followed in my wake, and before they could answer, I began to hear several young women and children singing in the distance. Poignard was standing in the middle of the road with a curious expression on his face. He must have run out in a hurry, too, because he was clutching a steaming cooking spoon. I followed his gaze and saw a large crowd of teenage girls and children approaching us from the north.

Aristote caught up to me and said, "A baby was born in the village today, and now the women are coming to celebrate with us."

As the crowd and the singing drew closer, I could see several of the smiling girls leading the chorus, and one of them had a stick that she was banging against a plastic container like a drum. At the front of the group, four girls, ranging in age from eight to twelve, were dressed in traditional attire that included speckles of white body paint on their limbs and faces, skirts and matching tops that seemed to have been constructed from large banana leaves that had been cut into ribbon-like strips, and matching cone-like hats that were adorned with flowers and fastened with bamboo head straps. Immediately recognizing the stars of the show, but again ignoring common courtesy, Poignard ushered the decorated girls forward while holding the rest of the crowd back. Two of them grasped little twigs as if they were wands, and all of them held up their hands and wiggled their fingers like they were playing an imaginary piano. They wiggled their shoulders and hips to the beat of the music.

"*Oh doh deeyay!*" they sang.

Unable to restrain his energy, Bakota flung himself into a never-ending series of cartwheels. Minutes later a second wave of dancing teenage girls arrived, one of them wearing the cone-like hat, while two of the others donned the vegetation in a way that reminded me of green dreadlocks. Instead of the white flecks, orangish-red pigment adorned their faces. They wiggled their shoulders and swayed back and forth, kicking their left feet with a flourish with each lurching movement. A group of older girls, some of them carrying babies, danced and sang in rhythm at the back of the crowd. At the end of each verse, some of them raised their arms or gently tapped the person next to them in excitement. The dance resembled a swaying march, forward and back, with the occasional dip while shaking the shoulders.

I noticed an ancient, toothless woman in a blue headwrap at the back. At first, she just observed the festivities with quiet approval, but slowly, she started to bop her head to the beat and then joined the girls in their dance and song. One of the girls was not moving quite right, so the old woman grabbed her from behind, and forced her body back and forth to show her the correct tempo. Taking this cue, another woman in her late thirties, clutching a baby to her hip, started poking and yelling at the girls from behind, obviously criticizing their technique.

The crowd increased until it seemed like the entire village had turned out for the performance. Teenage boys showed up with an empty jerry can and tapped on it with branches. In the finale, the eldest women pushed their way to the front of the crowd, presumably to show everyone how the dance should be done. All of them wore colorful, faded skirts and headwraps, and they waved sticks, reeds, and large manioc roots around as they danced and sang. The oldest woman, center stage, gave a solo performance whereby her hands and feet were on the ground, almost in a plank position, and she moved her limbs in a strange way that reminded me of a crab. At one point when she seemed to have trouble standing back up, a young boy brought her a walking stick, and she righted herself. The other women shook their accessories at her as they moved in unison toward her, seemingly singing and talking to her simultaneously.

I was so thunderstruck by the unexpected show that it took me several minutes to gather my thoughts and expand my awareness beyond the performers. Poignard, wrapped in his brown winter coat in a lawn chair on the sidelines, also seemed paralyzed with wonder. All the dancers had been facing me, but when I turned around, nobody was behind me. Did I somehow end up as the guest of honor? I realized that the Batwa were sharing a joyous celebration with us because they felt free enough to do so. It was only possible because we were in a village that did not have other tribes that would deride such cultural expression. I knew I was witnessing traditions that were likely thousands of years old, and time would prove that it was to be a once-in-a-lifetime experience.

Thanks to the Batwa, a steady trickle of frogs, snakes, and turtles from Lake Tumba and Npenda kept me busy in the laboratory for most of my visit, but I did not miss a chance to search the forest whenever I had time. Just after dusk one evening, following a particularly delicious meal from Poignard, we set out for a swamp that the Batwa had shown to Chifundera earlier in the day. The cloudless sky was full of stars, and I walked slowly as I tried to make out some of the constellations.

When we entered the forest, the terrain seemed somewhat similar to Mpote-Emange, because many of the trees had twisting labyrinths of roots that sprawled across the leaf-strewn terrain. The trees seemed to be dominated by different species, but I could not be sure—Congo has hundreds of species of trees, some of which look so similar that experts can distinguish them only by their smell or sap color.[81] Numerous kinds of palms were interspersed in the vegetation, and lianas were draped over the tree branches like giant pieces of spaghetti. Somehow the place felt different for other reasons too. At first, I thought it was the noticeably cooler temperature—certainly unusual, but that was not it. And then it hit me. It was eerily quiet with only a single lonely insect droning off in the distance—usually the forest was full of insect noise. I looked back to see glimpses of Wandege's light shining up into a tree, searching for animals in its tangly branches. The forest was so thick that the canopy completely entwined its branches and leaves, obscuring any view of the night sky. Everyone seemed to move quietly, almost with reverence, as we searched in vain for animals.

When Wandege caught up to me, I asked him, "Wandege, have you ever seen a place like this?"

His eyes slowly followed the beam of light from his headlamp as he scanned the surreal landscape. "No," he replied. "Nevah."

After a few more minutes of searching, we came to the edge of the swamp that the Batwa had told us about. A mist crept across it, obscuring my view beyond a few feet, and I noticed that a milky film had accumulated at the surface of the stagnant water. Given the lack of frog calls, we were definitely not there at the right time, but I shined my light into the water to see if I could catch the wriggling movements of tadpoles. Aside from some aquatic insects, I did not notice anything. I stared at the bottom of the water and saw a decaying mess of leaves, branches, mud, and unidentifiable remains of countless plants and animals. At that moment, I did not realize the significance of the seemingly lifeless morass in front of me, and I turned back, convinced that we had stumbled upon an inexplicable dead zone in the forest.

But now I know better, thanks to groundbreaking research on Central Africa's extensive peatlands. Also known as bogs, fens, and mires, peatlands are essentially waterlogged quagmires of dead plant detritus that are typically in acidic, nutrient-poor, and anaerobic (lacking oxygen) conditions, all of which slow decomposition. Much of the carbon that the plants absorbed from the atmosphere in life is thus stored in the peat. The resulting soil is like a sponge because it contains a lot of empty pore space; it is light and compact when dry, but it expands and becomes heavier as it absorbs water. In general, peat accumulation occurs very slowly, and radiocarbon estimates of Congo's peatlands suggest they began forming over ten thousand years ago.[82] Although peatlands cover only about 3 percent of Earth's surface, they store more carbon than all the other planet's vegetation types combined (including all forests), providing a net-cooling effect that mitigates climate change. Peatlands also help to maintain biodiversity, decrease the risk of floods and droughts, preserve air quality, and ensure safe drinking water. Damaged peatlands, often resulting from draining, agricultural conversion, and mining, are responsible for nearly 5 percent of global carbon dioxide emissions, but long-term restoration efforts can reduce these greenhouse gases.[83]

The importance of peatlands in Congo and neighboring countries was not fully appreciated until a 2017 study in the prestigious journal *Nature* brought it to the world's attention. Over fifty-five thousand square miles of interconnected peatlands were discovered in a relatively flat region known as the Cuvette Centrale, including the western edges of the Congo River, proximate tributaries, and nearby lakes, more than any other region in the tropics. The peat stored an estimated thirty billion tons of carbon, roughly equivalent to twenty years of fossil fuel emissions in the United States.[84] A subsequent 2022 study by the same scientists and several additional African colleagues increased the estimated area of Central Africa's peatlands by 15 percent, including peat swamps near Lake Mai-Ndombe where I had found the dwarf clawed frogs (*Hymenochirus*) at Mpote-Emange. Compared with other areas of the world, these peatlands are relatively undisturbed, but they will not remain that way for long. Only about 8 percent of this peat occurs in protected areas, some with stronger law enforcement practices than others, and many greedy interests threaten the rest, including exploration for petroleum and natural gas, logging, hydroelectric dams, and palm oil plantations.[85] Congo's muddy peat swamps are about as far away as one can get from innovative solutions to climate change, but we need them just as much as electric vehicles and solar panels. It is imperative that we do everything within our power to keep Congo's peatlands intact to ensure they do not become "climate bombs."[86] The hellscape that would result from the loss of Congo's peatlands is unthinkable.

WILD GREEN YONDER

Part of me did not want to leave
Npenda, but as amazing as our work had been, it was time to move on to other places in search of the new species mother lode. The Batwa watched stoically as we packed up our tents and loaded them into another geriatric truck that Chifundera had found in Bikoro. Somebody had resurrected the vehicle from the dead, because key components were missing, including my seat belt, the door and window handles, the side mirrors, most of the dashboard, and a huge amount of material that had once covered the steel frame. In the enclosed space I smelled something rancid and became paranoid that it was me—it had been a few days since I had a proper shower and I looked forward to one in Mbandaka, a major city on the Congo River just a few hours' drive to the north. As usual, several people had to help us kickstart the truck, and when it finally started, it was incredibly loud because the muffler was either long gone or so decrepit that it was useless. As we started to roll, or more accurately, limp down the road, the Batwa were smiling and waving. Bakota ran alongside us on the grassy shoulder until we picked up speed and showered him in a cloud of reddish dust as we passed him. I felt a little heartbroken to leave our new friends behind, but I am confident they will remember us if we ever return.

The jungle became increasingly sparse as we headed north until it petered away to small clumps of trees at the edge of Mbandaka. As we entered the city, I noticed quaint Belgian houses with distinctive arches, chimneys, and carports, and entire city blocks with storefronts, churches, factories, and other major buildings. Sprinkled throughout the city were foreigners, including South Asian shopkeepers, European UN workers (dealing with the aftermath of a civil war in nearby Central African Republic), and Chinese businessmen. Looking at the large font size associated with Mbandaka on my map of Congo, by far the largest city in Équateur Province, I had high hopes we were heading into a metropolis with amenities that would give us a break from the privations we had endured since leaving Kinshasa. But as I would come to learn, Mbandaka's location at the edge of the wilderness, together with its colorful and tragic history, resembles small towns of the Wild West more than a modern African city.

The precursor to the city was founded in 1883 as Equator Station by Henry Stanley during his initiative to establish a series of stations along the Congo for his patron King Leopold II.[1] Stanley left Belgian Lieutenant Camille Coquilhat in charge.[2] Soon thereafter, the station became renowned as a trading post for ivory, and Stanley's medical officer Thomas Parke remarked that he traded a teaspoon of salt for an elephant tusk that might fetch windfall profits of 200 percent.[3] The station was abandoned in 1886, but after Belgium assumed control of Congo in 1906, Coquilhatville was established just upstream from the original station in the "angle formed by the left bank of the Ruki" and Congo rivers.[4]

In the twentieth century, Coquilhatville (nicknamed Coq by the locals[5]) established racially segregated schools, hospitals, and housing, a colonial pattern that was repeated throughout Congo. But at certain times and places within the city, these boundaries seemingly evaporated. "At night, in nervous Coquilhatville as in smaller towns, pleasure, fashion and desire ruled. When white, black, and Portuguese crossed in Congolese bars, they drank beer, danced to love songs among other rumba tunes, and took in the femme libre garb, smooth and flashing."[6] Prostitution was sanctioned by the state, and in 1950, the governor of the province actually levied a tax of 150 francs for "femmes libres" (free women).[7]

Mobutu spent his formative years with an uncle in Coq, where he attended a school run by Belgian priests. He made his classmates giggle by lobbing ink darts at the backs of the teachers while they worked on the blackboard.[8] By the early 1960s as an increasing number of Europeans fled the Belgian Congo, the city was derided as "an inhospitable backwater, a sleepy toe-hold of civilization in the midst of a vast jungle."[9] In the 1970s, as Mobutu's "Zaireanisation" (i.e., nationalization) policies caused an economic collapse,[10] an attempt was made to create jobs in Mbandaka with a Bralima brewery (a subsidiary of Heineken) and a Chinese agricultural mission briefly attempted to grow rice in the marshes surrounding the city.[11] The darkest chapter occurred during the twilight of Mobutu's roughly three-decade rule. In the aftermath of the Rwandan genocide in 1997, hundreds of desperate Hutu refugees fled over one thousand punishing miles from their homeland to Mbandaka, where they hoped to find a way to cross the Congo River into the safety of Republic of the Congo. But they had been pursued by the vengeful Rwandan army (Tutsi) and their Congolese allies, who trapped and slaughtered them, including women and children.[12]

At the height of the colonial era in the mid-1950s, Coquilhatville had an estimated 1,000 Europeans and 30,000 Congolese.[13] At the time of my visit in 2013, the mostly Congolese population of Mbandaka had ballooned to about 352,000.[14] Although the city had a working airport and docks for the barges and ships that stop on the water highway to Kinshasa or Kisangani, the easternmost city on the navigable stretch of the Congo River, I soon realized that Mbandaka is more a large town than a city. The best hotel we could find (the Benghazi) had no internet connection or any cybercafe nearby, so we settled for the Centre d'Accueil Epervier (Sparrowhawk Reception Center). It did not have running water or air conditioning, but the power worked most of the time, and a floor fan kept us cool during the hottest parts of the day. The local restaurants were delicious but expensive, so Poignard set up a grilling station in the hotel's courtyard and continued to cook for us while we resupplied and planned the next leg of our journey. A portable radio suddenly appeared at his side, and from that day on, his cooking was always accompanied by blaring rumba music, the wildly popular genre throughout Congo. I thought about paying Poignard off and looking for another cook, but as I discussed the issue with Chifundera over some grilled chicken and carrots, he sighed.

"I know Poignard is . . ." said Chifundera as he paused to search for the right word, "a ruffian."

"Yes!" I vehemently agreed.

"But if we dismiss him, I don't know how long it will take to find a replacement."

"I understand," I said. It was true that we hoped to resupply, find a boat, and head east into the wilderness as soon as possible.

He licked his thumb in a place where the chicken juices had dribbled, looked at me with a knowing expression, and asked, "And do you *really* want to give up this food? Nobody else can cook gourmet like this."

"No," I admitted. "But if we continue with him as our cook, we will be too remote to dismiss him later if we have a problem."

Chifundera smiled and said, "Let us leave it to God."

And so we did, even though I could not shake the feeling that I was taking a huge gamble. Then again, the entire expedition was riddled with so many risks that one more did not seem to matter. I was keen to work in Salonga National Park, Congo's most remote park, about 188 miles east of Mbandaka, at least as the crow flies. But months before when I had traced the path of our proposed journey on Google Earth, the corkscrew appearance of the rivers revealed that the true distance was at least twice as long. As we would head east, the size of the rivers would decrease too—we would travel from the mile-wide Congo, a major artery, to the tiny capillary of the Yenge—only a few dozen feet from one bank to the other in places. Getting stranded there, miles away from the nearest village, would be a potentially life-threatening catastrophe. But big risks come with rich rewards.

The park was first conceived in 1957 when King Leopold III of Belgium, then president of l'Institut pour la Recherche Scientifique en Afrique Centrale (IRSAC), flew over the Salonga, Yenge, and Loile rivers. Apparently wanting a closer look, he then traveled southeast by boat from Watsi Kengo via the Salonga River, where he counted two hundred forest elephants along the riverbank. He also explored the Yenge in a smaller boat, where he saw the beautiful and unspoiled forest, and decided a large park should be established to protect it. Plans were underway to do this, including a large fund to relocate people living in the proposed park, but Congo's independence delayed the project.[15]

Salonga National Park was created by Mobutu in November 1970, simultaneously with other parks in eastern Congo, including Maiko, Kahuzi-Biega, and Kundelungu.[16] Unlike the latter three parks and most other protected areas in Central Africa, Salonga was (and still is) distinguished by its vast size (about 14,000 square miles in two large sections), remoteness (accessible only by boat or plane), minimal deforestation, and low human population density.[17] Because of these factors, its status as Africa's largest protected tropical rainforest, and the presence of rare and endangered species like bonobos, forest elephants, and the Congo peafowl (*Afropavo congensis*), Salonga was inscribed as a UNSECO (United Nations Educational, Scientific, and Cultural Organization) World Heritage Site in 1984.[18] But as political instability and conflict spread in Congo in 1999, widespread poaching, habitat destruction, and water pollution prompted Salonga to be included on the List of World Heritage in Danger sites.[19]

With the exception of outstanding long-term studies on bonobos and ecological investigations of birds and relatively large mammals,[20] most of Salonga is a zoological terra incognita, because its remoteness has discouraged biodiversity surveys since its creation; what little is known is often buried in hard-to-find reports or remains among personal field notes.[21] In a way, this is a scientific tragedy because the knowledge must be subjected to rigorous quality control (i.e., peer review) and widely disseminated in a reputable scientific journal to be accepted and useful to the scientific community.

In the case of amphibians and reptiles, Belgian herpetologists were mostly focused on surveys of national parks in eastern Congo during the colonial era, and very few collections came from the remote region that would later become Salonga.[22] When I searched for "reptile" and "Salonga" in Zoological Record, the gold standard database for zoological research, I found nothing. There was a little more for amphibians, but it barely scratched the surface. Schiøtz spent "a few days in and around" Salonga in 1975, including Watsi Kengo, and he published a small paper that discussed two species of spiny reed frogs (*Afrixalus*).[23] Decades later, his protégé Jos Kielgast visited Salonga in January 2008, and soon thereafter, he published two studies with colleagues. One short note documented foam-nest frogs (*Chiromantis rufescens*) using empty weaver-bird nests to deposit their eggs, possibly as a way to hide them from hungry monkeys.[24] The second described a handsome new species of reed frog, *Hyperolius veithi*, with a distinctive yellowish stripe from the tip of the snout along the flanks to the sacral region, from forests of the southern section of Salonga.[25] Obviously, the herpetological hole of knowledge at Salonga was enormous, and it was a distinctive opportunity to fill at least some of the gap. Moreover, given Salonga's inclusion on the List of World Heritage in Danger sites, I also felt an urgency to see the park while it was still relatively pristine, even if my limited resources would allow me to visit only a tiny corner of the park's northwestern edge. But getting there would be a treacherous logistic challenge.

When it came to finding a boat, fuel, and a trustworthy captain, I relied on Chifundera. It took him a few days of searching, but he finally found a doctor who agreed to let us rent his dugout canoe and two outboard motors (as an emergency backup in case the first one died in the middle of nowhere) for twenty days at the low-low price of $800. I paid more than double this price for about three dozen jerry cans full of gasoline to take us to Salonga and back, because we would be so remote that it would probably be impossible to find any fuel along the way.

All of us were happy to abandon the stifling heat of the Sparrowhawk on July 14, and we soon found ourselves at a beautiful colonial-era house with a grassy backyard that overlooked the Ruki River. Pieces of the siding and paint were flaking away from decades of neglect, but otherwise, the house looked much as I imagined it did in the early 1960s when it was surely abandoned by its Belgian owner in the violent aftermath of independence.[26] An enormous mango tree monopolized the western flank of the property, and I noticed several other strategically planted lemon trees nearby. Ancient rusting speedboats were discarded along the edge of a fence, and I craned my neck to look at a fifty-foot-tall television antenna that was still supported by thick wires, which were anchored into slabs of fissured concrete on the ground.

When I walked to the northern edge of the lawn I saw a veritable cliff, and twenty feet below it, the bank of the Ruki. Poignard, Aristote, and Wandege were already busy shuttling our gear and supplics down an attractive staircase made from rocks, which snaked its way down the steep slope from the backyard to a dock at the river. There I noticed our pirogue, festooned with a bright orange tarpaulin canopy to provide shade and shelter from the rain, just as I had requested, and a tall man loading our gear into it. Looking downriver, I could see several other similar houses with the same vantage of the river and the opposite deforested bank. A gentle breeze drifted up from the river, and for a fleeting moment, I relaxed my eyes and drifted into meditation.

I felt slightly guilty about enjoying the tranquil beauty of the vista, because it had been created within a framework of injustice, racism, cruelty, and a condescending ethos of paternalism. Like many Congolese towns and cities in the colonial era, Coquilhatville had been designed to shield the white colonial elite from "contacts inutiles" (useless contacts) with the much larger and poorer Congolese majority. Separating these two worlds were undeveloped tracts of land occupied by camps of the notoriously barbaric Force Publique, who enforced a nighttime curfew to ensure the wealthy European neighborhoods would be white only. In places where the Belgians prospered and expanded, the Congolese neighborhoods were pushed to the periphery of the city, often to places with subpar resources and unhealthy environments.[27] Was it really that different, I wondered, from the treatment of African Americans in my own country?

The loud crack of someone dropping a bag of cooking charcoal onto the rocky courtyard jarred me back to reality. I joined the team, carried some materials down to the boat, and introduced myself to the tall man who turned out to be our captain. About six feet, two inches tall and with a build like a professional basketball player, Jonathan had gentle eyes and a calm disposition. He seemed to have a knack for organization, and I felt confident that we were in good hands. Everything seemed to be packed and in order, but then Chifundera said we had to wait for immigration officials from the dreaded DGM to arrive before we could leave. I sighed in frustration as I recalled the madness of Maluku. This time we were at a private dock, and until now, I had been under the mistaken belief that we could slip away without them noticing. But I should have known better—a white man in Africa is never invisible.

For an hour, I returned to the scenic overlook, lounged in the grass, and watched a few boats filled with firewood and other goods drift by on their way to the Ruki's western merger with the Congo River. Finally, an unusually corpulent man, often a hallmark of corruption in Congo, arrived on a motorcycle, wiped the copious sweat from his brow, and demanded to see my passport and the official documents authorizing our mission. A few minutes later, his skinny assistant arrived on foot, licked his lips, and jumped into the animated interrogation. I watched in silence under the shade of the mango tree, regarding the DGM men as a Laurel and Hardy comedic duo. For every query, Chifundera had either a reasonable response or an official document, and the minutes drifted by as the officials scrutinized the stamps that peppered each scrap of paper. Realizing they were running out of options to create a big problem, the assistant glanced at me and suggested I should stay in Mbandaka so that I could teach a course in English. It took more time to explain that I already had a job to do and it would actually benefit his country if I could accomplish it. In the end, Chifundera paid them $15 for nautical maps we did not need, and everyone was relieved to see them go.

For a moment I relished the quiet sounds of waves on the river that resumed after the DGM grilling faded away. I glanced at the team and saw Jonathan, Aristote, and Wandege stretching their legs now that the spectator sport had ended. And then I noticed Poignard was missing.

When I asked Aristote about our cook, he shrugged and said, "When he heard DGM was coming, he left very quickly."

Shocked, I asked, "Well, did he say when he was coming back?"

"No," he replied.

I cursed and turned my attention to Chifundera, who was now engaged in a conversation with Jonathan. Both of them had concerned looks on their faces and I knew more bad news was coming. Chifundera slowly walked over to me, his gaze toward the ground as he scratched his head. He told me that one of the boat motors was not working properly and Jonathan was off to look for a replacement. Moreover, the boat I had already paid to rent was too big, and he would need to find a replacement for that too. I tried to stifle my exasperation as I questioned why Chifundera, the expedition logistics leader, had failed to look into all of this before we were ready to leave.

"I understand your perspective," he began. "But in Congo things are *very* difficult . . ." He continued to tell me things I already knew about Congo's widespread problems with unreliable equipment and people, unforeseen circumstances, corruption, scarce resources, and other woes until I told him not to bother explaining more. I was frustrated, but we had been working together for years, and both of us knew these setbacks were a fact of life in Congo. Somebody was already trying to find a solution, and there was nothing to do except wait. The perennial Congolese gift to me was a huge dose of patience. I found a comfortable place in the grass, opened a book, and tried to lose myself in its words.

As the hours drifted by and our stomachs started to grumble, I sent Aristote to town to buy us grilled chicken for lunch, and he returned with Poignard, who acted like it was perfectly normal to slip away without telling anyone. When I questioned him about this, he said he needed new batteries for his radio. I caught Wandege smirking at his answer, and all of us knew it was bullshit. I frowned at him and told him that if he wanted to continue working for us he should not do that again.

"Ok, boss, no parabalem," he replied in broken English that he had picked up from Aristote. His overall demeanor was deferential, but there was an annoyed and angry expression on his face that was impossible to mask.

There was no sign of Jonathan as the sun started to wane in the late afternoon, and acknowledging defeat, we pitched our tents on the lawn and settled in for the night. By way of tacit apology, Poignard obtained and cooked spicy fish and potatoes, and I fell asleep trying to think of our situation with the "glass half-full" perspective.

Dawn arrived with miraculous news—we had our replacement motor and boat. We wolfed down a quick breakfast of eggs and scrambled to transfer everything to our new ride. By nine a.m., we were seated in lawn chairs under the tarpaulin canopy of the pirogue and motoring east along the Ruki. We passed some crumbling ports, complete with rusting cranes, as well as some old

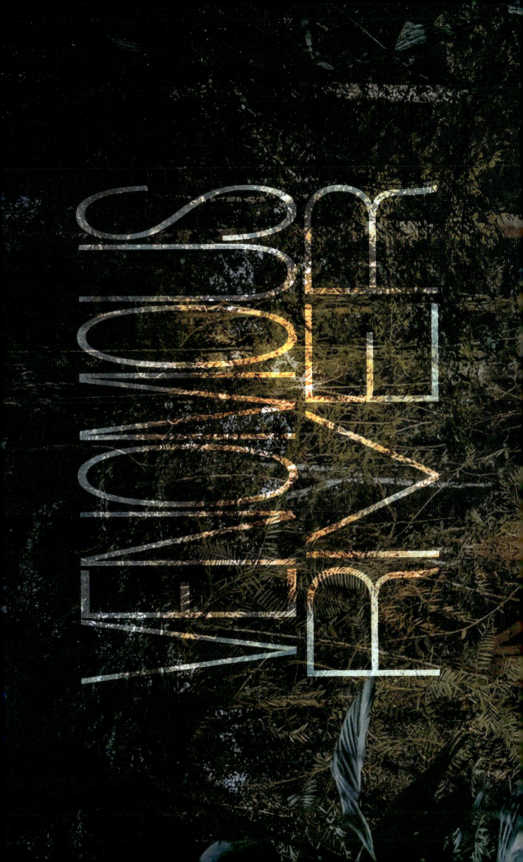

The variable burrowing asp (*Atractaspis irregularis*) that nearly ended Wandege's life. Photo by the author.

A branch-and-liana bridge over the Lumene River. Photo by Chifundera Kusamba.

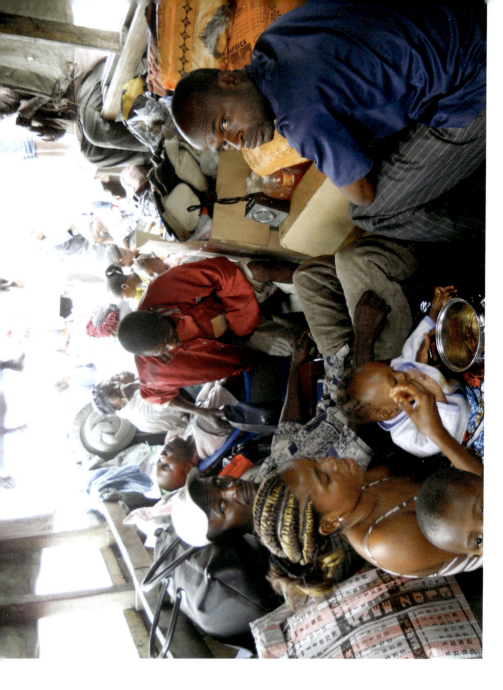

Day 2 on the cramped boat from Maluku, Kinshasa. Photo by Wandege M. Muninga.

An adult male chimpanzee (*Pan troglodytes*) performs a vocal display from a tree in Budongo Forest, Uganda. Photo by Adrian Soldati.

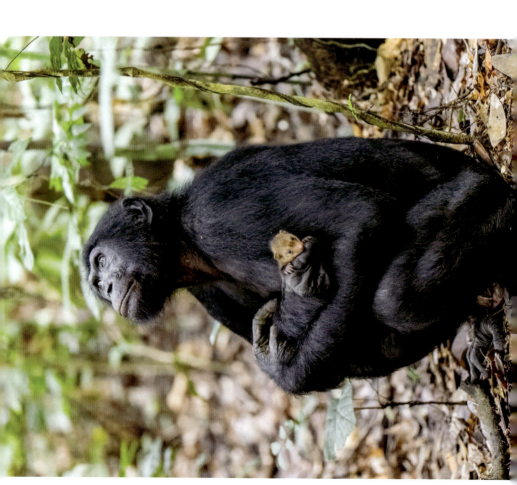

This award-winning photo by a *National Geographic* photographer shows a young male bonobo (*Pan paniscus*) holding a mongoose pup at LuiKotale field station near Salonga National Park, Congo. Photo by Christian Ziegler.

(*below*) A *Gerrhosaurus* lizard found at Kempili Bembe. Photo by the author.

Stilt roots in the seasonally flooded forest at Lake Mai-Ndombe. Photo by the author.

An enormous fig tree (*Ficus*) at Mpote-Emange village near Lake Mai-Ndombe. Hammer-headed fruit bats (*Hypsignathus monstrosus*) called from its branches every night. Photo by the author.

A golden puddle frog (*Phrynobatrachus auritus*) found in the forest at Mpote-Emange, near Lake Mai-Ndombe. Photo by the author.

A banded water cobra (*Naja annulata*) found at the Salonga River near Bomputu. Photo by the author.

Wandege searching for water cobras with a young fisherman in Lake Mai-Ndombe. Photo by Mwenebatu M. Aristote.

(*below*) A dwarf water cobra (*Naja nana*) with a banded pattern from Lake Mai-Ndombe. Identification confirmed with morphological data. Photo by the author.

Aristote gets hunting lessons from the Batwa of Isongo. If you look closely, you can see he is holding an inflorescence of the African locust-bean tree (*Parkia bicolor*), the "microphone" to God. Photo by Wandege M. Muninga.

Hogtied dwarf crocodiles (*Osteolaemus osborni*) at Isongo. Photo by Mwenebatu M. Aristote.

(*above*) Chifundera entertains the Batwa of Npenda with the playback function of his camera. Photo by Mwenebatu M. Aristote.

A likely new species of African gliding lizard (*Holaspis*), captured by Batwa near Lake Tumba. When threatened by predators, these lizards can leap from 150-foot-tall rainforest trees and use fringes of skin to glide safely to the ground. Photo by the author.

(*above*) A white-bellied pangolin (*Phataginus tricuspis*) captured by Batwa at Npenda. Photo by Mwen-ebatu M. Aristote.

Batwa girls at Npenda prepare to dance in celebratory costumes. Photo by the author.

A lone fisherman paddles his dugout canoe on the Salonga River. Photo by the author.

(*below*) A likely new species of forest treefrog (*Leptopelis*) from Ikakao near Bomputu. Photo by the author.

Aristote chats with children near our tents and towering clumps of bamboo at Watsi Kengo. Photo by the author.

(*below*) A rare burrowing cobra (*Naja multifasciata*) found at Watsi Kengo. Photo by the author.

The beautifully camouflaged and potentially deadly rhinoceros viper (*Bitis nasicornis*). Photo by the author.

A juvenile slender-snouted crocodile (*Mecistops leptorhynchus*). Photo by the author.

Wandege dons a winter hat in our dugout boat as he looks at the forest along the Yenge River. Poignard keeps watch in the bow. Photo by the author.

(*below*) A powdered tree snake from Salonga National Park that was named as a new species (*Toxicodryas adamantea*) by the author and colleagues in 2021. This species is venomous, but no snakebite cases have been recorded and the toxicity of the venom remains unknown. Photo by the author.

The herpetology team pauses a daytime search for animals in the forest at Lotulo, Salonga National Park. Pictured from left to right are park ranger Jean-Louis Lokwa Bonjulu, Chifundera, the author, Aristote, and park ranger Wema Engulu. Photo by Wandege M. Muninga.

A splendid tree snake (*Rhamnophis aethiopissa*) found at Efofa on the Salonga River. Photo by the author.

A venomous green bush viper (*Atheris squamigera*) flicks its tongue at Watsi Kengo. Photo by the author.

(*below*) A young Central African rock python (*Python sebae*) found near the Salonga River. Photo by the author.

The one-horned *"unicornis"* variant of *Trioceros oweni*. Photo by the author.

churches, warehouses, and colonial-era mansions with waterfront views. Just beyond the eastern edge of the city, we entered a huge marsh that seemed to be carpeted in water hyacinth and other thick vegetation. Fishermen had built little wooden bamboo and thatch huts on stilts above the water. I could not imagine why someone would purposely want to live in a mosquito-ridden swamp with no shade from the powerful equatorial sun, but perhaps they wanted to be close to the market where they could sell their daily catch. Or, more likely, abject poverty gave them no choice.

Slowly, almost imperceptibly, the palms and small trees on the banks transitioned into incrementally larger trees until we were finally surrounded by a seemingly pristine rainforest on both sides of the river. Fishing eagles (*Icthyophaga vocifer*) with distinctive white heads and white-tipped wings became increasingly common, as did various other birds. White butterflies with a reticulate black pattern on their wings were suddenly everywhere, and in some places, dozens of them fluttered around the water adjacent to our boat, occasionally nosediving into the water for a quick drink. It reminded me of a scene with dancing fairies from the movie *Fantasia*. At other times, beautiful lime-green or vibrant-blue butterflies darted in between us in the boat, keeping up with us for half a mile or more, seemingly curious about us. Although we traversed long stretches of riverine forest without seeing another human being, we passed by tiny fishing villages on the riverbank every few miles or so. We thus started a pattern whereby we would camp in one of the friendly villages just before dusk, share some of our food and money in thanks, and then pick up and leave the next morning.

After two days we reached the Busira River and stopped at a small clearing on the northern bank called Balolombo, a Batwa village that seemed to survive on a healthy combination of hunting, gathering, and fishing. In the evening, a light thunderstorm commenced, providing ideal conditions for frogs, and the Batwa offered to guide us to a small stream where they said we could find them. In our haste and excitement, we left without our rain jackets and followed our guides into the forest. The rain coalesced through leaves of the thick canopy until it combined into heavy droplets that plunked onto our bodies with audible impacts, bursting like tiny water balloons. Within minutes we were soaked and a little cold, but when the Batwa led us to the magical place in the forest that they had told us about, nothing else mattered.

The stream, puddles on the ground, vines, branches at waist level, and the trees above our heads were crawling with frogs of every conceivable size, shape, and color. Brown and gray clawed frogs (*Xenopus*) and dwarf clawed frogs (*Hymenochirus*) wriggled in the puddles. Golden puddle frogs (*Phrynobatrachus auritus*), some distinguished by chocolate brown and tan backs, others with lime-green snouts and orange limbs, hopped along the edge of the stream. Orange-brown fishing frogs (*Aubria*) with black and white polka dot bellies hunted in pools formed by the rain. Rusty red and mustard-yellow forest toads (*Sclerophrys*), obviously new species, hopped in the leaf litter. Calling forest treefrogs (*Leptopelis*) with tan and orange bodies and orangish eyes emitted "*Quaaaaaack*" sounds from every available perch overhanging the stream. Green-eyed frogs (*Cryptothylax greshoffii*) with orange bodies and metallic eyes perched on large leaves. Lime-green reed frogs (*Hyperolius*) with golden yellow borders to their snouts shared some of the leaf real estate. White-lipped frogs (*Hylarana*) evaded capture with powerful jumps until Wandege made a daring dive to catch one, soaking himself at the edge of the stream in the process.

In the morning, as I had just started the hours-long process of photographing all the frogs we had found the previous night, fishermen brought us water snakes (*Grayia*) and turtles (*Pelusios*) that had become ensnared in their nets. The place was so productive that we decided to stay longer than planned. On the second morning, we found a day gecko with a leopard-like pattern that eventually proved to be a new species—it was dubbed *Lygodactylus leopardinus* in 2023.[28]

On the third day, we left Balolombo in a light rain and headed east, eventually crossing into Tshuapa Province and reaching the Salonga River. Along the way, for countless hours, we rarely saw any people. But at one point, out of the blue, I noticed a pirogue similar to mine with a white man in the back of it. He did not have the benefit of a canopy to shade himself from the punishing tropical sun, and a white towel was draped over his crimson, sweat-soaked head.

When he heard the drone of our outboard motor, he lifted his head long enough to see me, jerk his neck back in shock, and wave. We were moving in opposite directions and were too far away to speak, but I waved back and wondered what he was doing out here in the middle of nowhere. He surely had the same thought.

Later we waved to a dozen people who were drifting downriver on a large raft constructed with logs and a tarpaulin suspended by a spine-like branch for shade. Two pirogues were lashed to the side of it, and a young boy was steering with an oar from the stern. The women yelled and waved their arms for gifts, but we passed each other too quickly to exchange anything. I noticed a pot-bellied boy on the bow, probably about five years old, his slightly older sister, and a pregnant teenage girl.

"Look!" yelled Aristote as he pointed at the rickety structure with a disdainful smirk. "That is their house! Their toilet and bath is the river! They are going to Mbandaka to sell fish."

"Isn't that dangerous?" I asked, concerned for their safety.

"Yes, of course," he replied with a shrug. "But they must find a way to make some money."

Astonished by the desperation and poverty, I watched as the family and their precarious craft disappeared behind a bend in the river. Aside from these brief glimpses of fellow travelers on the river, our journey east was mostly devoid of other human encounters, but I never really tired of looking at the forest. No stretch of it was the same—the jumbled mix of tree species was extraordinary. In some places along the edge of the river, tiny reeds transitioned to rattan palms and then enormous giant trees over one hundred feet tall. Here and there one could spot even taller, lone giants that had truly vanquished their neighbors in the race to reach the sun. Unimpeded by competitors at the same height, their uppermost branches spread out in round crowns that resembled giant stalks of broccoli. It was as if I was staring at enormous green walls on each side of the river, and only the tiny fishing villages interrupted them. The palisade of trees, palms, lianas, and other vegetation was so dense that one could see only a few feet into the dark forest. In this place, humanity seemed ephemeral, and I had never felt so diminished and insignificant.

In the late afternoon, we reached a sizable village called Bomputu. It looked like about one hundred people lived among two dozen simple buildings, and we decided to pass the night there so that we could look for chickens, fresh vegetables, and other groceries to restock Poignard's cooking pot. I found a spot at the edge of the village to pitch my tent in the shade of some enormous clumps of bamboo that towered at least thirty feet above me. A fine carpet of dead leaves covered the ground underneath them, and because I had the usual distracting audience of excited and curious children around me, I almost failed to notice a small, yellowish-brown scorpion that was crawling on the ground near the edge of my tent, a well-camouflaged needle in a haystack of similar-shaped leaves.

Later it would be identified as the lesser brown scorpion (*Isometrus maculatus*),[29] a species with one of the largest geographical distributions of any scorpion. Never exceeding three inches in length, the species likely originated in tropical Asia, and hitchhiked its way through much of the warmer parts of the world via colonial-era shipping routes.[30] Its penchant for hiding under pieces of wood and detritus probably facilitated its spread across the globe, even to places as remote as Bomputu.[31] The venom is relatively mild and usually causes only pain, numbness, and weakness at the site of the sting.[32]

I was used to the pattern of arriving at a remote village where white visitors are rare, politely greeting the chief and village elders, asking for permission to stay for a day or two, and having throngs of children and a few curious adults crowd around me as I set up camp. But the screams of the excited children who watched me capture the scorpion attracted more people to the crowd surrounding me until half the village seemed to be there. When the animated chittering did not cease after two hours of patiently waiting, I retreated to my tent, but then they resorted to gawking at me through the tiny window in the rain fly like I was a caged animal in a zoo. They jostled around the window until I resorted to covering it from the inside with duct tape.

The murmuring crowd had just started to disperse when a woman's terrible wailing welled up from the center of the village. Distracted by a more entertaining spectacle, I heard most of them leave, and I stepped out to investigate. I caught up with Chifundera, who explained that an eight-year-old girl had died of some unknown disease. Apparently, Bomputu had a "health center," but Chifundera explained they had no medication or trained nurse, and in reality, it was little more than a comfortable place to die if someone became ill. It was a chilling reminder of the risky journey we were taking. Even a minor mishap could become deadly without medical care.

At dusk as we searched the forest at the edge of the village, I soon discovered one of the possible sources of disease at Bomputu. Despite its relatively large size, nobody had bothered to dig a latrine anywhere, and the periphery of the village was recognizable by its ring of human waste. I suggested that they should change the name of the village to Bompoopoo. I was also reminded of Jean-Pierre Hallet, a linebacker-sized Belgian who spent his formative years in Congo, and his astute observation that, "Africa can indeed be a place of vast, almost overwhelming beauty, but it is also a place of mud, blood, and excrement."[33]

But even here, oblivious to the stench, forest treefrogs (*Leptopelis*) were calling for mates from waist-high perches on palms with enormous thorns, and given their small size and distinctive orange and brown coloration, it seemed that we had discovered another new species. Some of them had lemon-yellow borders at the edge of their limbs. Their irises were truly remarkable—silvery gray with black vermiculations except for the upper quarter, which resembled the orangish-red color of sunset. The entire eyes were bordered by a thick black ring, giving the appearance of heavy mascara. I managed to record the call from one of them, but we had to be extremely careful to avoid impaling our hands and limbs on the enormous palm thorns that the frogs straddled while calling, perhaps as a way to avoid predators.

Congo's popular rumba music greeted us as we returned to our tents. Even here in the middle of nowhere, someone had a speaker and enough power to generate music, and they used it to blast the same twenty songs for twenty hours per day. Even with earplugs, sleep was impossible until the party ended at four a.m..

In the morning as I shared some coffee with Chifundera I asked, "Don't they get tired of listening to the same songs? Don't the neighbors complain when they play the music until four a.m.?"

He smiled and shook his head in response. Apparently not. When I suggested we might move on, Chifundera said we needed more time to resupply. I would figure out later that the real reason for our extended sojourn in Bomputu was that everyone but me had searched for and found female company and nobody was in a hurry to leave. Sensing my restlessness and learning that good food could appease me, Poignard brought my attention to the village bakery, a true rarity in remote areas of Congo. I purchased fresh baked rolls for everyone, relished them with chocolate and peanut butter packets, and had to admit that Bomputu had its charms.

Wandege helped me photograph frogs as Chifundera and Aristote went to investigate a nearby Batwa village called Ilala. They returned in the afternoon with a charismatic young Batwa man named Bola, who happened to kill a rare tree cobra (*Pseudohaje goldii*) the day before our arrival. Luckily, he had not eaten the snake yet, and we salvaged what we could from it, including enough muscle tissue for a solid DNA sample. I showed my gratitude by sharing some of our food and other supplies with Bola, and he did not need much convincing to help us search for more animals.

Just before dusk as we were preparing for another foray into the forest and a welcome reprieve from the rumba music, Chifundera approached me with a pained expression. Bola had captured an extremely rare *Polemon* snake, unmistakable by the bright orange ring around its neck, but as he was approaching Bomputu to bring it to us, he was stopped by a policeman. For reasons that were impossible to rationally understand, the cop beat Bola, and during the incident the container with the snake fell to the ground and it escaped. I was angered by the violence and concerned for Bola. Chifundera assured me that Bola's injuries were minor and he would talk to the policeman to ensure that it did not happen again, at least while we were around.

When I suggested we should report the policeman, Chifundera's eyes grew wide, and he asked, "Report to who? He is the only law here."

I was shocked by the injustice of the incident and disheartened by the reality that I was powerless to mitigate the centuries-old mistreatment of Indigenous people, even for one person. I vowed that in the future, to ensure their safety and dignity, we would only collaborate with the Batwa when there was no possibility of interference from other tribes.

Following another sleepless night from the rumba music, I had just finished a cup of coffee when Chifundera unexpectedly showed up with Bola and the policeman to smooth things over. The latter man had clearly dressed for the occasion, because his dark blue uniform and black beret were clean and pressed, and his black boots were shined. A thick black baton swung from its fastening to his belt. When I looked at Bola, I did not see any visible injuries, but in sharp contrast to my first meeting, his demeanor was subdued, his eyes were affixed to his bare feet, and he was clearly traumatized to be near his antagonizer.

The middle-aged policeman had a slightly misshapen face, suggesting he had been on the receiving end of a beating himself once, and his eyes looked sunken and black, giving him a corpselike appearance. As he approached, he glanced at Bola, looked at me, and cracked a ghoulish grin. Tacitly, his smile seemed to seek my approval. My drowsy body did not have enough time to react to this horrifying behavior before he reached out his hand to shake mine, but something in my expression conveyed my displeasure. Chastened and nervous, he stole glances from me as Chifundera explained that Bola was helping us with our mission and that we needed his cooperation. The policeman nodded, Chifundera lightened the mood with some small talk, and somehow, I found the reserve to say "Merci" as we parted ways with one of the most vile human beings I have ever encountered.

Bola slowly returned to the cheerful man we had met the day before when we shared our food and coffee with him, and we did our best to reassure him that the policeman would leave him alone—at least for the rest of our visit. Bolstered by the support, Bola guided Chif and Wandege into the forest while I worked with Aristote. Hours later when we reunited, Chifundera told me that they saw one of Congo's most elusive reptiles—a keel-bellied lizard (*Gastropholis*). Nearly impossible to catch, these roughly foot-long, vibrant green lizards have prehensile tails that are often twice as long as the body, which might be used to traverse flimsy vegetation such as thin branches or stems. They live in hollow trees far above the forest floor, probably spending most of their time in the canopy, and even when they venture lower, they are extremely alert and agile.[34] During the American Museum Congo Expedition (1909–1915), one of the few specimens known from Congo was obtained by pure luck when one of the local tribes felled a tree to cross a stream and snatched the shell-shocked lizard after the impact. Another one was shot with a bow and arrow.[35]

In the afternoon as we listened to strange howling from the village dogs, Bola returned from his search of the forest near Ilala with a handful of frogs. One of them was a reed frog with a coloration I had never seen before. It was olive green with yellowish blotches on the back of its tailbone and some of the leg articulations; thick, yellowish stripes extended from the middle of the flanks over the eyes, to the tip of the snout where they conjoined. Once back in my lab, and with the benefit of its DNA, I was able to confirm that it was *Hyperolius veithi*, the first specimen documented outside Salonga.[36]

Poignard joined us with some news. The bakery had run out of flour, and with it, the biggest attraction in Bomputu, at least from my perspective. But on the positive side, he said somebody had brought in fresh meat, and he could cook it for us. Knowing that the meat could only have come from a wild animal, I was just about to inquire about the source when Aristote ran up to us, wide-eyed and breathless.

"I can't believe what I just saw," he began. "Did you hear the cries of the dogs?"

"Yes, why?" I asked.

"The people called one of the dogs over to them, but they were hiding a big stick. When the dog came, they killed it with the stick and now they are selling the meat! The other dogs watched it happen and now they are crying."

All our jaws dropped, except for Poignard. Chifundera glared at him and asked the question that was on everyone's mind.

"Est-ce la viande que vous vouliez cuisiner?" (Is this the meat you wanted to cook?), he asked in disbelief.

Poignard instantly knew we objected, so he shrugged his shoulders and quietly replied, "oui."

"Oh!" Chif exclaimed with a disapproving glare.

Traumatized, Aristote repeatedly shook his head and said, "I can't understand. How can they do that to their own dogs?"

Nobody said anything for a minute as we listened to the mournful howling.

"Chif, I have had enough of Bomputu," I said. "We are leaving tomorrow."

This time, there were no excuses to procrastinate.

7
CROCODILE CITY

The twenty-seven currently recognized species of crocodilians (crocs for short, including living crocodiles, alligators, caimans, and gharials) are included among the roughly twelve thousand species of reptiles,[1] but in many ways, crocs have more in common with other groups of animals, both living and extinct. Archosaurs, an ancient group of animals dating back at least 245 million years ago (emerging shortly after the Permian–Triassic extinction event), shared a suite of esoteric osteological characters (e.g., possession of a fourth trochanter on the femur bone) that demonstrated a close relationship between early crocodilians, birds, pterosaurs (extinct flying reptiles), and dinosaurs.[2] In fact, some scientists hypothesize that by understanding crocodilian ecology, behavior, and physiology, we can obtain a window into similar features of dinosaurs.[3] The earliest recognizable crocs in the Triassic (252–201 million years ago, mya) were relatively small terrestrial predators with long legs—the opposite of modern crocs, but in the Jurassic period (201–145 mya) new marine crocs (with flippers instead of legs) evolved, along with other species that more closely resembled modern crocodiles.[4] By the early Cretaceous (145–66 mya), the roughly forty-foot-long, 17,600 pound "super croc" *Sarcosuchus imperator* appeared and, given that it was slightly larger than a *Tyrannosaurus rex*, it likely preyed on dinosaurs.[5] Fossils of *Sarcosuchus* have been found in Africa and South America because these continents, originally conjoined in the supercontinent Gondwanaland, had not yet drifted far apart at the time of the super croc's heyday.[6]

Although crocodilians and birds, the surviving archosaur lineages alive today, are starkly different in many aspects of their physical appearance, physiology, ecology, and behavior, they retain some key similarities that evince their ancient, common ancestry. Crocodiles have an inner ear structure that is remarkably more bird-like than any other animal; it can be used to convert sound waves into electrical impulses that are used for hearing, but it is also utilized for balance and position. In contrast to the three-chambered heart of most reptiles, crocodilians have a four-chambered heart like birds and mammals, and because of long periods of time submerged underwater (sometimes up to seven hours), crocs have what is arguably the most complex heart of any vertebrate. Like birds, crocodilians also have a unidirectional flow of air within their lungs.[7]

Birds and crocs both have surprisingly diverse vocal repertoires, but their distinct evolutionary paths have forged different ways of producing sounds. Songbirds vocalize via the syrinx, a complex structure located at the base of the windpipe (i.e., trachea) with a cartilaginous skeleton and tympanic membranes that produce sound when air passes over them during expiration. Syringeal muscles control tension on the membranes, which in turn affects the pitch of the sounds.[8] Many birds can produce sounds in other ways, too, by rapidly banging their radius wing bones together (manakins) or using deep wingbeats (trainbearers) to produce snaps, or by utilizing their beaks (woodpeckers) or sticks (palm cockatoos) like drumsticks on resonant surfaces.[9]

Crocodilians and other reptiles do not have a larynx (i.e., voice box), their vocal cords are rudimentary, and muscles associated with the lungs do not have the ability to affect vibrations of the vocal cords in the same way that mammals can. And yet, crocs are considered to be the most vocal reptiles. Observations of Australian saltwater crocodiles (*Crocodylus porosus*) documented at least a dozen categories of sounds, including roars, growls, bellows, and even infrasound vibrations at low frequencies that are below the range of human hearing. Some of the sounds happen during performance art via "narial geysering" (blowing water fountains out of the nostrils), snout water slaps, jaw claps, raising open jaws out of the water while vocalizing, and blowing bubbles underwater, probably to get the attention of mates.[10] A breeding pair of captive individuals of the West

African slender-snouted crocodile (*Mecistops cataphractus*) were recorded giving a duet of roars and growls.[11]

Most of us are familiar with the dedicated parenting of birds, typically involving brooding behavior (keeping the eggs and nestlings warm), shuttling back and forth to the nest with food, and the performance of distraction displays to lure predators away from nests.[12] With the exception of some turtles and a tiny fraction of egg-laying snakes and lizards that usually have more limited brooding behavior of eggs or nests, the vast majority of reptiles have no parental care to speak of, but not so with crocodilians. Depending on the croc species, both sexes can be involved in parental care, but females typically shoulder the largest share of the responsibility, often staying near nests and defending it against potential predators. When the juveniles are ready to hatch from the eggs, they will begin to vocalize within them,[13] attracting at least one of the parents to assist them by gently excavating the eggs, breaking them open, and carrying the young ones to the water. Depending on the species, one or both of the parents will then stay with the babies for several weeks to a year or more, sometimes sharing food with them, and distress calls by the youngsters will trigger defensive behavior by the parents.[14]

Crocodilians are carnivores, and they will ambush and eat increasingly larger invertebrate and vertebrate prey as they grow, including fish, amphibians, reptiles, birds, and mammals. Several species of larger crocs are known to eat humans that venture into their realm, but they are not always the sinister monsters many of us might suspect them to be. Captive individuals can be trained to recognize their names, respond to commands, and submit to medical checkups.[15] A Costa Rican man named Chito became famous for nursing an American crocodile (*Crocodylus acutus*) with a gunshot wound back to health in his house (much to the chagrin of his wife), and forming a bond with the one-thousand-pound creature that he named Pocho. Chito swam with Pocho for twenty years, sometimes on his back, often posing with the seemingly tame animal for disbelieving tourists.[16] Watching documentary footage of their intimate interactions, along with Pocho's suspicious behavior toward the filmmaker, it is impossible to dismiss the crocodile's intelligence.[17] A 2023 observational study of Mugger crocodiles (*Crocodylus palustris*) in India documented the species' affinity for colorful marigold flowers, strategic use of bait for hunting birds, cooperative fishing, and a completely unexpected interaction with a dog. Although the crocodiles are known to ambush and eat dogs, a young stray that had wandered into the wrong territory was chased by a feral pack of dogs into the Savitri River where three adult Mugger crocs were lurking. Two of them immediately approached the dog, but instead of attacking it, they nudged it with their snouts to the opposite bank where it escaped. Why would they do this? Although wildly speculative, the authors suggested it could be cross-species emotional empathy.[18] No matter how one interprets these sophisticated behaviors, it is clear that crocodiles are not mindless killing machines, and considering all the evidence, they seem to be sentient.

But just as we are beginning to understand them better, many crocodilians have been pushed to the brink of extinction. Unregulated hunting of millions of crocs for food and skins in the nineteenth and twentieth centuries decimated their numbers across the world until the IUCN/Species Survival Commission (SSC) Crocodile Specialist Group and CITES were established in 1971 and 1975, respectively. Their resulting regulations and management plans allowed some species to slowly recover.[19] Since then, there have been some notable success stories, with the American alligator (*Alligator mississippiensis*) rebounding so well that it was removed from the federal endangered species list in 1987.[20] Zoo captive breeding programs of other croc species have produced offspring that were released into the wild as a key component of in situ conservation efforts. But thanks to habitat loss, illegal hunting, pollution, invasive species, mismanagement of water resources, and other human-caused problems, other croc species have not fared as well. Seven species are critically endangered, the most serious level of extinction threat, including the gharial (*Gavialis gangeticus*), which suffered a 90 percent population decline and roughly 94 percent reduction in its area of occurrence in India, Nepal, and Bangladesh over the last seventy-five years.[21]

Until recently, Africa was known to have three species of crocs in three genera, including the Nile crocodile (*Crocodylus niloticus*), slender-snouted crocodile (*Mecistops cataphractus*), and the dwarf crocodile (*Osteolaemus tetraspis*). All three were recently studied in DNA-based phylogenetic analyses, and in each case, at least one cryptic species was present. In other words, what was thought to be three widespread species of crocs turned out to be seven with more limited geographic distributions, and conservation assessments have not yet caught up with this updated taxonomy. One study identified the likely mechanism that separated West and Central African populations of *Mecistops* and *Osteolaemus*—uplift of the Cameroon Volcanic Line, a series of volcanoes and highlands that formed in the mid- to late Miocene, roughly simultaneously with the estimated speciation timing of the crocs.[22] In the case of *Osteolaemus*, the smallest of the true crocodiles, genetic analyses suggested three species are present—an unnamed one in West Africa (as far east as Ghana), *O. tetraspis* in the western part of Central Africa (including Gabon, Cameroon, and neighboring countries, west to Togo), and *O. osborni* in Congo.[23] Because CITES regulations create red tape that makes scientific collection of crocodile specimens extraordinarily difficult, genetic sampling of the African populations is relatively sparse, and the exact boundaries between these species remain fuzzy.[24]

Africa's most iconic crocodile has been likened to a dragon, an apt comparison given its enormous size and reputation as an irascible man-eater. Indeed, people are attacked and killed by Nile crocs on a regular basis in Africa, but we are just hors d'oeuvres. Reaching an impressive eighteen feet and two thousand pounds, and sometimes even more, Nile crocs regularly ambush, kill, and eat massive prey, including large antelope like wildebeest, giraffes, zebra, and even rhinos.[25] The study that split Nile crocodile populations into two distinct species is especially intriguing. Using DNA from samples collected all over Africa, including mummified Egyptian crocs from Thebes and Samoun that are as old as 200 BCE, scientists reconstructed their evolutionary history and karyotypes (chromosome profiles) and found that two cryptic lineages of Nile crocs were present (*Crocodylus niloticus* and *C. suchus*), and in fact, they were not even each other's closest relatives. Although it has now disappeared from Egypt, the mummified samples and historical accounts from ancient Egypt suggest the smaller, less testy species (*Crocodylus suchus*) was preferred in temples and ceremonies. Both species were known to occur in northeastern Congo in the early part of the twentieth century, but understanding of their current geographic distributions and conservation threats in Central Africa is incomplete at best.[26]

The slender-snouted Crocodile can reach lengths up to thirteen feet, and it was recently split into two species, *Mecistops cataphractus* in West Africa and *M. leptorhynchus* in Central Africa.[27] A fair amount is known about the ecology, biology, and behavior of the former species, but the latter one has been dubbed the "least studied crocodylian in the world." We do know that the Central African species seems to favor fish for its diet, but it will also eat insects, crustaceans, snakes, and even water chevrotain (*Hyemoschus aquaticus*), a twenty-pound, deer-like mammal. Female slender-snouted crocodiles construct mound-shaped nests behind screens of vegetation, suggesting a reclusive nature when nesting, and they lay thirteen to twenty-one eggs. The species vocalizes frequently during the mating season, perhaps to attract mates or demarcate territory, and its call has been likened to a "low growl" that is similar to a lion. Like other crocodiles in Africa, it has likely suffered massive declines throughout its once expansive range, but given the lack of recent surveys, nobody can be certain about population sizes or where all the remaining strongholds are located.[28] Given the complete absence of any wild crocodile sightings along our route from Kinshasa to Bomputu, I knew it would be a long shot to catch even a brief glimpse of them.

On a bright, steamy afternoon on July 24, we rounded one of the endless bends of the Salonga River and arrived at a makeshift port where people, motorcycles, and goods were shuttling from a road on one side of the river to the other via pirogues and a wooden ferry. After weeks of passing through small fishing villages and long stretches of jungle without seeing another soul, the sudden commotion of many people was jarring. Motorcycles backfired as they struggled to ascend the steep road leading to Watsi Kengo, goats bleated as their owners violently tugged on rope leashes, and chickens lashed to luggage by their feet shrieked and shed feathers as they flapped their wings in vain attempts to escape. Mothers wrapped in colorful, sweat-soaked dresses from the afternoon heat ignored crying babies and toddlers as they gossiped with friends on the muddy bank.

Chifundera, notebook in hand, gestured to a passing man and initiated a conversation. Two minutes later, having obtained the requisite intelligence, he looked at me and said, "Come with me, we must meet with the authorities." Not wanting to miss an opportunity to make new friends, Aristote joined us, and Poignard did, too, after mentioning he needed to look for food. I glanced at our cook and noticed that he was walking with an awkward gait, as if he was sore from riding a horse. He avoided eye contact, seemingly self-conscious about his discomfort, and I made a mental note to check on him later.

Leaving Wandege behind to keep an eye on our gear, we ascended the slippery road and entered the town of Watsi Kengo, where I saw hundreds of people going about their business, dozens of houses and buildings, and several small shops selling everything from cheap imported goods to bread. Distracted by an attractive woman selling cassava, Poignard stood up straight in an attempt to hide his pain and drifted away toward her.

Aristote watched Poignard greet the woman, grinned, and said, "Maybe I can meet a woman here too. They are pretty."

As we continued toward the center of town we took in the bustling scene of people and goods moving in all directions. Impressed by the size and amenities of the town, I commented, "Wow, these goods have come a long way from Mbandaka."

"Maybe, but probably they came from Boende," Chifundera replied with a gesture toward the north. "They have Lebanese traders there, and planes bring goods from Kinshasa."

Boende is a city with an airstrip on the Tshuapa River about forty-five miles northeast of Watsi Kengo. It definitely seemed more feasible to transport goods from there than the long river journey from Mbandaka. For a moment I contemplated the difficulty and isolation someone would face by moving from Lebanon to a river outpost in Congo. And then Chifundera snapped me back to our current situation.

"We are actually lucky that Boende is not too far away, because Jonathan said we need more fuel for the boat."

"Wait . . . what?" I asked.

"Yes!" Chifundera continued with a judgmental glare, as if I should have psychically known that we were running low. "He miscalculated how much fuel we need to reach Salonga, and we will need to buy more in Boende if we want to reach Salonga and return to Mbandaka." To emphasize his point, he raised his hand and folded a finger for each of the responsible factors. "You must know that fuel is affected by size of the boat, number of passengers, baggage, wind, river current, and what I can say . . . many other things."

Shocked, I stopped walking, stared as his folded fingers now crumpled into a fist, took a deep breath, and inquired about the exorbitant cost that would surely be required to send Jonathan on a motorcycle to Boende and back with two jerry cans of fuel. Chifundera gave me the ballpark number, I winced, briefly considered objecting, realized there was no alternative, sighed, and continued walking.

"Oui," I said in a defeated tone.

Chif gave me a knowing smile, tacitly acknowledging my adaptation to the Congolese modus operandi. Aristote grinned and gave me an encouraging slap on the shoulder. It still sucked.

Following a few minutes of asking around, we were introduced to a handsome man named Antoine in his early thirties who said he was the deputy mayor of Watsi Kengo. Chifundera started explaining our mission in the middle of the street, but Antoine gestured to a bar with a canopy where he preferred to discuss our business in the shade. Aristote's eyes lit up as Antoine ordered a large bottle of Primus beer, confident that I would be willing to pay for it. Several greasy glasses were distributed on the table as we enjoyed the cold beer and discussed our plans to spend a few days in town before going on to Salonga. Antoine's beautiful girlfriend showed up to say hello and help us down more beer. Four or five bottles later, we were the best of friends and our host assured us he would give us a safe and comfortable area to pitch our tents for a few days.

I paid the bill and tried not to stumble as we stood up to leave. Aristote sighed contentedly as we retraced our steps with Antoine until he directed us down a narrow path toward the forest. At least a dozen curious men, women, and children followed us to see what we were up to. We reached a dell of sorts that was sprinkled with waist-high vegetation and surrounded by towering clumps of bamboo that seemed to be fifty feet high. Resembling giant bunches of parsley, the bamboo blocked out most of the sun, creating a shady space for our next camp. Delighted, we thanked Antoine and started to make our way back to the boat to retrieve our gear, all the while followed by the group of onlookers with nothing better to do.

When we reached the port, we found Wandege with a large, wriggling snake bag. He explained that while he was waiting for us, he spotted a large snake that had descended from a nearby tree in search of the chickens that were wandering around near the bank. Wandege opened the bag and showed us a jet-black snake that inflated its neck in a threatening display. The onlookers shrieked in horror and questioned how we could handle such a large serpent without being killed. It turned out to be a magnificent, five-foot-long specimen of the black tree snake (*Thrasops jacksonii*), a harmless species that is known to ascend one hundred feet or more into trees in a large area of Central and East Africa. They eat a variety of animals that can be found in trees, including chameleons, bats, birds and their eggs, but they will sometimes leap out of trees to attack terrestrial prey like frogs.[29] We were encouraged by the lucky find, and word spread quickly that we were looking for snakes.

We returned to the bamboo campsite, luggage in hand, to find Poignard and two women using machetes to clear away the vegetation in the center of the dell for our tents. I recognized one of them as the woman who had been selling cassava. After we set up our tents, Chifundera approached me to ask whether it would be possible to hire the women as assistants to help Poignard with chores for firewood, water, laundry, and cooking. I glanced over at him and saw that the women were already helping him cook our lunch.

Chifundera noticed my skeptical expression and said, "Yes, maybe he likes one of the women, but it can be helpful to hire them for a couple of days to show the local people that we are good."

I considered the pros and cons. Ostensibly, Poignard did not really need any help for his chores, but maybe his limp was caused by an injury, and hiring a couple of locals would indeed demonstrate our good intentions to the people of Watsi Kengo. Although he was a great cook, Poignard's laundry skills were terrible, and it would be good to get our clothes properly washed, at least for now. Even with the unforeseen expenses of obtaining fuel in Boende, I had enough financial cushion to hire the women, so I agreed.

In the evening we searched the forest near the edge of town and found several frogs, but they seemed to be species we had found previously, and we decided to search in more pristine forest the following day. I was sleeping soundly when I was awakened by people singing and drumming from somewhere on the other side of Watsi Kengo. I looked at my watch—three a.m. Sleep was impossible until they stopped just before dawn.

Awakened by the creation of a cooking fire from Poignard and his coworkers an hour later, a small rivulet of sweat trickled down my temple as I raised my head from my pillow. The humidity seemed to be thicker than usual here, and my clothes stuck to my body with a clammy tenacity. Gloomy light trickled through the stalks of bamboo and when I emerged from my tent, the atmosphere resembled dawn more than early morning.

My vision was still adjusting to the light when I noticed a blur of motion just inches from my tent. Focusing on the movement, I saw the telltale hops of a toad, and lunged forward to grab it while managing to keep my bare feet inside the tent. The damp leaf litter soaked through my shirt and pants in seconds, but I managed to keep my feet dry. Switching on my headlamp to get a better look at the four-inch-long frog, I saw that I had captured a Cameroon toad (*Sclerophrys camerunensis*), a bit of a misnomer given that it was originally described from Nigeria.[30] Like other species of forest toads, this one had a series of cone-shaped tubercles from the back of its mouth along the flanks, and its body was colored with a mixture of brown, tan, and cream, including a white stripe down the middle of its back and a pair of dark brown blotches between its eyes, all seemingly designed to resemble leaf litter and fungus on the forest floor. Most of the forest toad species I had encountered on the expedition were completely new to science, and with subsequent study, I would learn that nearly all of them had very restricted geographic distributions, often hemmed in by small rivers that they did not have the ability to cross. By contrast, the Cameroon toad occurs across most of the forested regions of Central Africa, and genetic analyses of the specimens I caught confirmed the species easily crosses from one major ecoregion to another, including biogeographic barriers like the Congo River that are insurmountable obstacles to many other species of vertebrates.[31] Why this one species of toad is the equivalent of a special forces soldier when the others are not remains a mystery for now.

Aristote smiled when he noticed me sprawled in front of my tent, admiring the elaborate camouflage pattern of the toad, and he greeted me with a steaming cup of tea in his hand. Jealous, I carefully washed my hands to ensure all of the toad's poison from its paratoid gland was gone, and then fished around in my backpack for a Via coffee packet before giving Poignard some of my purified water to boil. Amused, Aristote and I watched as he informed the giggling women that the *mondele* can die without coffee, not *really* an exaggeration, and that he follows this process every morning to avoid such a catastrophe.

One of the women donned a red bandanna and seemed to be in her late thirties, and when she noticed me looking at them, she stopped laughing and stared at me with a hardened, calculating glare.

"Mondele! Donne-moi un cadeau" (White man! Give me a present), she yelled with a flippant gesture.

Normally I would respond positively to such requests, but I had just given her a job, and her tone was demanding and unjustified. Instinctually, I recoiled and could not hide my expression of shock. I turned to Aristote who was watching our exchange from the sidelines.

"Aristote, what do you think—should I give her a present?"

"Noooo," he said emphatically. "Maybe she can have a gift if she does a good job, but after."

She eyed him suspiciously as he uttered the "No" to me. I agreed and asked him to pass along the sentiment in Lingala. I thought that would be the end of the matter, but I could not help feeling uncomfortable every time she glanced at me as she worked with Poignard at his makeshift kitchen. I could not explain it rationally, but there was another feeling too—dread. I decided Poignard would work alone on our return trip to Watsi Kengo, but everyone would stay on until our departure. I returned my attention to Aristote.

"Did you hear the drums last night?" he asked.

"How could I not?" I responded. "Were they having a party?"

"No!" he responded. "A woman and her baby died in childbirth last night, and many people came for the mourning ceremony. Now they must be suffering."

Yet another sad reminder that in this part of Congo, basic health care is out of reach for many. Suddenly my sleep-deprived headache seemed trivial and my reluctance to hire the extra women now struck me as selfish. If I had the resources, and I could make someone's life a little bit better for a time, why not do it?

We had barely finished breakfast of rice and river fish when a fisherman showed up with a live water cobra (*Naja annulata*) that had turned up as bycatch in one of his nets. Normally he would have sold it as meat for the equivalent of a dollar, but wanting to be more generous with the people of Watsi Kengo, I paid double and encouraged him to come back if he found any more. Within an hour, more people showed up to direct us to places where someone had killed a snake recently, and I had my hands full photographing specimens that Aristote and Wandege retrieved from different areas of the town. With so many eyes peeled for snakes, we even managed to track down snakes that were still alive, because people noticed them crawl under piles of firewood or other rubbish at the edge of the forest, and someone would keep watch as one of us hastened to capture it before it wandered off.

In this way we found an impressive bounty of snakes, including a green snake (*Philothamnus*) with an odd beak-like snout, a *Polemon* with a thick gold collar that extended onto half of its head, and a spectacular rhinoceros viper (*Bitis nasicornis*) with a kaleidoscope of colors. Rhinoceros vipers reach lengths of about 4.5 feet and are relatively common in lowland rainforests of Central Africa, where they spend a lot of time in motionless ambush of mammal or amphibian prey.[32] The venom is known to cause potentially serious swelling and coagulopathy (uncontrolled bleeding), and an American hobbyist died from anoxic brain damage after receiving a bite from a captive animal.[33]

In the afternoon a group of boys showed up with a harmless brown skink known as a tropical mabuya (*Trachylepis polytropis*), which they had chased down in the forest. I had never encountered the species before, and I was struck by its unusually large ten-inch size, noticeably larger than the size of the more common speckle-lipped mabuya, and its yellowish throat and green belly. Years later, genetic analyses would suggest that this animal was so genetically distinct that it likely represents a new species to science.[34] A similar pattern would be found for several species of frogs that we found at Watsi Kengo, including a squeaker with orange legs and a yellow belly (*Athroleptis*), forest treefrogs (*Leptopelis*), and a spiny forest toad (*Sclerophrys*) that looked like someone had applied neon orange lipstick along the edge of its upper jaw. It was so distinctive that it took only a quick glance to realize it was a new species.

Just when I thought things could not get any better, one of the rarest snakes in Congo showed up. Somebody had spotted a snake slowly crawling over a fallen tree in the forest, and when Wandege managed to capture it, he could not believe his eyes. About a foot long, the serpent was mostly black with creamy yellow scales around the face and lower parts of the flanks, and similarly colored spaces in between scales of the back gave it the appearance of jagged bands. It was a burrowing cobra (*Naja multifasciata*), one of Central Africa's most secretive snakes. Rarely reaching 2.5 feet in total length, this cobra seems to spend most of its time sheltering out of sight, and thus, next to nothing is known about its ecology, behavior, and natural history.[35] Given its small size, the neurotoxic venom may not be deadly to humans, but mice have succumbed relatively quickly to bites from captive specimens, and we handled it with extreme caution.[36]

When Jonathan finally returned from Boende with the fuel we needed a couple of days later, all of us were somewhat reluctant to leave, because the collecting had been fantastic and the amenities of the town seemed almost luxurious compared with those of the small fishing villages we had frequented since leaving Mbandaka. But on the twenty-eighth of July, excited by the prospect of sampling a completely unknown corner of Salonga National Park, we left Watsi Kengo and headed east, stopping only briefly in the tiny village of Efofa to buy an enormous catfish that Poignard spotted hanging from a tree near someone's hut. His limp was barely perceptible as he ascended the muddy bank, paid the fisherman after a brief negotiation, and shuffled back to the pirogue in a celebratory dance with the glistening prize, accompanied by a small cloud of flies, promising to cook the fish with the best *pili pili* spices that he had procured in Watsi Kengo.

Wandege smiled and said, "*Quel gourmand!*" (What a food lover!).

The excitement of our future dinner had died down by the time we reached a bifurcation in the river, and we took the southern fork, passing from the Salonga into the Yenge. Suddenly, the banks of the river were only a stone's throw from each other, and the water ranged from the color of strong tea to light brown. Small white sandbars flanked by elephant grass and palms were seen on every other bend of the river, and some of the rainforest trees clung to the banks at forty-five-degree angles, apparently desperate for a foothold among the crowded edge of the forest. The water was shallow enough to see fallen trees just below the surface, and we braced ourselves as some of the branches scraped the bottom of the pirogue, producing jarring vibrations. Our steady pace of travel suddenly became a drifting crawl as we maneuvered slowly and carefully to avoid colliding with trees hidden just under the surface of the water. When our outboard motor caught on one of the submerged branches, Jonathan cut the engine and pushed us past the obstacles with a long pole that he had obtained in Watsi Kengo.

Nobody said a word as we absorbed the prehistoric atmosphere in awed reverence. Until now, there had always been an expectation that we would pass a fellow traveler or two on the river, spot a lone fisherman tending his nets in a quiet backwater, or turn a bend to find a small fishing village. No more. Every tacit presence of humanity melted away as if it had always been ephemeral, especially during the frequent occasions when Jonathan had to resort to his pole to push us through some minor obstacle, with quiet splashes the only sound. A couple of times the submerged trees were so close to the surface that we had to get out of the pirogue and drag it across the branches, hoping that none of them would gouge a hole in the hull. The silence seemed oppressive, reminding us that we had never been so isolated in our lives, and if some minor catastrophe happened here, it might be weeks before another soul would happen to pass by to rescue us. The jungle did not seem to be fundamentally different from what we had seen on the Salonga River, but now we stared at it more intently, even fearfully, sensing that the ancient trees disapproved of our trespass into their ancient world.

Seven hours after leaving Watsi Kengo, we reached the remote park outpost of Lotulo,[37] where a handful of Salonga rangers clad in dark green uniforms welcomed us with waves and smiles. Happy to have company, they helped us unload our boat and brought us to a small clearing next to a *perruque* that they used as a kitchen. We pitched our tents, lingered in camp long enough to chat with the guards and share our catfish feast, and as the last inkling of afternoon light faded away, we set off down a narrow path into the forest. We were accompanied by a guard armed with an AK-47 everywhere we went to protect us from poachers, just in case. Scores of them had recently entered the park to slaughter the few remaining elephants for their ivory, because Chinese demand had sent the black market price soaring.[38] But in reality, we were not in any danger because the intelligent elephants had fled into extremely remote areas of the park, and the poachers surely knew to avoid Lotulo.

Within minutes, we found a reed frog (*Hyperolius*) that would eventually prove to be a new species, and Aristote spotted a powdered tree snake (*Toxicodryas adamantea*) with a unique orange coloration that was climbing a tree. The following day, our good luck continued as I searched with Wandege along a small stream in the forest. We found an impressive diversity of frogs, many of which would later prove to be new species, including a close relative of the golden puddle frog (*Phrynobatrachus auritus*), a white-lipped frog (*Hylarana*), a colorful forest toad (*Sclerophrys*), dwarf clawed frogs (*Hymenochirus*), and a rare spiny reed frog (*Afrixalus equatorialis*) with mustard-yellow toes. Wandege even managed to find a Gabon turtle (*Pelusios gabonensis*) that was swimming in the stream.

I worked with Aristote to set up a photography studio near our tents, but we soon realized that we had company. A hive of bees must have established themselves nearby, because every couple of minutes, one of the insects landed near us to investigate our equipment, clothes, sweaty skin, and anything else that caught their fancy. Everyone suffered a sting or two every day we remained in Lotulo, except for me, despite one intrepid bee who crawled up my pant leg. It escaped when I hastened to stand up, loosened my belt, and let my pants drop to my knees as I simultaneously cursed and begged the invader to spare my most sensitive anatomy from the stinger. All the guards laughed hysterically.

Aristote, who was stung twice, said it was unusually painful and lingering. The person who suffered the most was Poignard. The bees went crazy for the food around his cooking fire, and they stung him on his neck and face repeatedly as he tried to swat them away. During small breaks between the photo sessions and searching for animals, I watched as he limped around camp with his radio, pretending that his painful leg and swollen bee stings were a minor nuisance.

I caught up with Chifundera and asked him if he knew about Poignard's leg. He shook his head as he explained that Poignard had an infected wound that would not heal. Everyone insisted that he go to a clinic to get treatment, but for reasons nobody could understand, he refused to go. We speculated that it was tough-guy pride, but perhaps he wanted to keep a low profile to avoid unwanted attention. Regardless of the reason, Poignard's dismissal of our sound advice seemed to alienate him even further from us. As our expedition progressed, our understanding of him only worsened, and most of us found it impossible to connect with this human enigma. Only Aristote, beloved by all, had managed to befriend him, but even he could not explain Poignard's reluctance to care for himself.

A search of the forest that evening turned up absolutely nothing, a frustratingly common occurrence when conditions are not just right, but when we returned to our campsite, my attention was drawn to a clump of vegetation surrounding a large tree near the bank of the Yenge. A cacophony of tinks, buzzes, and cricket-like chirps originated from every hidden nook and cranny, suggesting at least two species of frogs were present. Donning my audio recorder, I set to work with Wandege and Aristote. Each of them would silently point to a leaf where one of the reed frogs (*Hyperolius*) was calling, and once I spotted the vantage point of the amorous male, I would start recording and turn off my headlamp to ensure it would not frighten the frog into silence or retreat. After a minute or two of successful recording, I would turn my headlamp back on, the tacit signal to capture the frog, and thus link the individual to its song.

After an hour of this process, we had recorded several frogs, but I did not recognize any of them. One individual was russet with lime-green flecks and golden yellow toes, another was dark brown with light green flecks and a thick cream band from the tip of its nose to its hip, and the largest individual was pinkish brown with black and green spots and a yellow belly. The latter individual, at least twice the size of the other reed frogs, bowed the uppermost stalk of the plant it was perched on with its weight. Although I had never seen one like it before, the strange black and green spotted pattern seemed strangely familiar. Then it hit me—although there were some minor differences, it seemed to be a close match to the mystery reed frog found by Schiøtz at Lake Tumba in 2006. Years later, and with the benefit of DNA analyses, I would confirm that this specimen was *Hyperolius sankuruensis*, a species previously known from a single locality over two hundred miles away.[39] Many rare species in Congo have poorly known geographic distributions, and only by piecing together localities of specimens like this, one lucky find at a time, can a clear picture of their range be surmised.[40]

I stepped closer to the side of the tree in an attempt to record the next frog spotted by Wandege, but Aristote put his hand on my shoulder and broke the silence of our routine by saying, "Careful, there are crocodiles in the water!"

Glancing behind me, I was alarmed to see that I had drifted within a foot of the riverbank, and a few feet beyond that, I could see the bright beacon of a *Mecistops* eyeshine in the water. Crocodiles have a crystalline layer behind the retina called the tapetum lucidum, which is present in many other nocturnal vertebrates. Its purpose is to reflect low light back to the retina to enhance their night vision—it gives them an eye "shine" that is salient in the beam of a flashlight.[41] The logical side of my brain calmly registered that slender-snouted crocodiles do not attack people, at least under normal circumstances, and I did not need to be concerned. But my instinctual consciousness fired a volley of fear and adrenaline into my body to warn me of the dragon-like threat lurking in the murky water nearby.

Pausing my efforts to record the frogs, I turned my body and headlamp toward the Yenge, scanning the water to see if the crocodile was alone. In fact, she had friends. Several glowing orbs reflected light back to me as I searched the small stretch of the river visible to me, including one individual who glided by in steely silence. Each of them seemed to be around six feet long, and aside from their eyes, the only visible parts of their bodies included the snout and head, designed by evolution with a low profile to avoid detection. I regarded them with amazement and wondered how many still survived within Salonga's borders.

On my last night at Lotulo, Aristote and I accepted an invitation from the rangers to look for crocodiles with spotlights. We piled into a dugout canoe and silently drifted downriver with the current, hearing many interesting calls and shrieks from unknown animals in the forest. Confused by our lights, fish jumped out of the water, sometimes into our boat. Attracted by the unfamiliar glow of our headlamps, moths flew into our faces, and stretches of the river seemed to be infested with biting insects. In a few hidden corners of the river, we could see the distinctive orange eyeshine of a crocodile, but they slipped underwater before we could get a really good look at them. Our jaunt down the river was cut short less than an hour later when a thunderstorm rolled in and soaked all of us in a downpour as we hastened to return to Lotulo.

I did not expect to see any crocodiles again when we said our goodbyes to the rangers of Lotulo the following morning and retraced our route along the Yenge. We were relieved to leave the bees behind, which disappeared shortly after our boat pulled away from the bank, but I was also sorry to part from the most pristine forest I had ever seen. We departed just after dawn, and the morning sun had just started to warm the earth along the steep riverbanks and sandbars that bordered seemingly every bend of the river. Wanting to warm their bodies from the evening's chill, the crocodiles had ascended from the water and were relaxing peacefully as they basked in the sun. Peacefully, that is, until the sound of our outboard motor and sight of our alien vessel rounded each bend to startle them. Their universal response was to take a running leap and catapult their dinosaurian bodies into the water. I will never forget the sight of crocodiles flinging themselves from the riverbanks as they came into view for a fleeting moment, and the sharp splash of water as they disappeared from sight. For at least an hour, we stared in astonishment as dozens of crocodiles flew into the water, and I knew I was witnessing behavior that few others would experience. It felt like we had entered a time machine to another world that pre-dated humans, where few predators would have dared to attack an adult crocodile.

As we left the boundary of Salonga and started our return journey to Watsi Kengo, all signs of the crocodiles disappeared. I was left to wonder how long the vast park (nearly fourteen thousand square miles[42]) could retain the remaining crocodile populations. When the flagrant and unsustainable poaching of crocs outside Salonga diminishes their numbers to the point where they are a rare delicacy, will the rangers of Salonga have the power to deter poaching within the park? The decimation of Salonga's elephants suggested a pessimistic outcome to me at the time, but in recent years the elephants have started to recover,[43] creating a wellspring of hope. Much more needs to be done to protect elephants across the African continent, including increased funding for conservation efforts.[44] The worst of human nature has precipitated the unfolding biodiversity crisis across the globe, and only unflinching resolve and support for parks like Salonga will protect the few wild places that remain.

PEEKING INTO PANDORA'S BOX

When Martinus Beijerinck
conducted experiments on a newly discovered disease that was decimating
tobacco plants in the late nineteenth century, he thought he was dealing with a
living microbe that he dubbed a *virus*, a word derived from Latin for a "poison,"
"venom," or "slimy fluid." Viruses can kill, cripple and maim, but they are not
alive, not really. On the one hand, they have scientific names like all of Earth's
organisms, including everything from bacteria to human beings. Viruses also
have their own genetic material and an evolutionary history that likely stretches
back billions of years, although they are far too small and fragile to have a fossil
record to know for sure. Viruses are the most diverse (circa 100 million types)
and abundant group of microbes on Earth, by far, and they occur everywhere
their host organisms are located, in every ecological niche possible, including
extreme environments like acid lakes and hydrothermal vents.[1]

But in sharp contrast to living organisms, viruses are not composed of
cells, and thus *in general*, they are not capable of making energy for themselves,
maintaining a steady internal environment (i.e., homeostasis), independent
reproduction, or most of the other things that are inextricably linked to life.
Viruses are obligate parasites in the sense that they need to hijack the infected
host's organelles and other cellular components to complete their life cycle. In
essence, viruses are tiny,[2] inert *particles* of protein and genetic material (either
RNA or DNA) that are programmed to invade cells, take control of their protein-
making components to replicate themselves, and move on to other hosts. Perhaps

immunologist Sir Peter Medawar said it best when he described viruses as "a piece of bad news wrapped up in a protein."[3]

The circumstances under which the first members of the virosphere (the gargantuan viral biomass on planet Earth) evolved and their relationship to living organisms is a hotly debated mystery, but viruses are capable of infecting organisms from every domain (Archaea, Bacteria, and Eukarya[4]), suggesting truly ancient origins, perhaps coinciding with the origins of life itself. Given this long history, it should not be surprising that fragments of viral genetic material have become inextricably entwined into the human genome over time. Our twenty-three chromosomes contain about 5–8 percent so-called endogenous viral elements (EVEs) that resulted from countless ancient infections whereby viruses permanently inserted their genetic information into those of our ancestors.[5] As everyone on planet Earth knows from the COVID-19 pandemic, new viruses are constantly popping up and trying to attack us, with some utilizing a more insidious strategy than others. For instance, it is blatantly obvious when someone has been infected with influenza—the telltale fever, body aches, fatigue, headache, and cough hit with rapid ferocity but usually subside after a few days. But other viruses like the human papillomavirus (HPV), one of the most common sexually transmitted infections in the United States (~80 percent of sexually active adults will become infected with the virus at some time in their lives), can hide in the body without symptoms for years before leading to potentially fatal cancers.[6]

One of the key features that distinguish major groups of viruses is the type of genetic material they contain, either DNA or RNA. The difference between the two forms is rather minor—DNA (deoxyribonucleic acid) has one less oxygen than RNA (ribonucleic acid) on the 2' carbon, making the former polymer less reactive and more chemically stable than the latter; this helps to explain why only DNA has been utilized as the molecule of heredity in living organisms.[7] But from the perspective of a virus, RNA has some key advantages. RNA viruses replicate quickly with many "offspring," and they lack the proofreading ability of other viruses and living organisms that have DNA as their genetic material. Thus, RNA viruses have a higher rate of mutation that can change their genomes more quickly. Mutations—heritable changes in the genetic sequence—are the random roll of the dice that fuel evolution. Many mutations have no effect, some are so negative that the virus or organism perishes, but some lucky mutations happen to be beneficial. Even a slight edge to hiding from the host's immune system, for example, can be helpful to the success of the virus strain that obtains it, allowing it to eventually outperform and outnumber older strains that lack the helpful mutation. This is why many RNA viruses rapidly evolve over a short period of time. Some of them, including ones like HIV, change so quickly that attempts to make vaccines have been extremely difficult, at least in the recent past.[8]

Some of the most notorious RNA viruses in the world have surprising connections to Congo. Human immunodeficiency virus (HIV), responsible for acquired immunodeficiency syndrome (AIDS), is thought to have originated from a similar virus carried by chimps in the Sangha River basin in Cameroon. The bushmeat trade, improved transportation networks, and establishment of colonial-era cities with concomitant factors like widespread prostitution and high population density eventually led to HIV's foothold at Léopoldville in the first two or three decades of the twentieth century. There it simmered mostly under the radar until colonialism collapsed in 1960 and the United Nations encouraged thousands of skilled professionals from other Francophone countries to travel to Congo to stabilize the chaos. At one point about ten thousand Haitians lived and worked in Kinshasa, and evidence suggests that one of them returned home and introduced HIV to Haiti, where it quickly spread, eventually ending up in other countries of the New World, including the United States in 1981. Kinshasa also served as a hub in a transportation network that allowed different variants of HIV to spread to other areas of Africa, and eventually, to the rest of the world.[9]

The most infamous virus in Congo, and arguably the most badass virus on the planet, is Ebola. Originally named for the Ebola River near the site of the first known *Zaire ebolavirus* outbreak at Yambuku, Congo, in 1976, the virus caused over three hundred cases and a fatality rate of 88 percent, which is probably higher than the rate for the Black Death bubonic plague in medieval France. The 1976 outbreak was exacerbated by the colonial-era, miserly practice of reusing syringes and needles after rinsing them in hot water.[10] Subsequent outbreaks averaged about 60 percent mortality rate,[11] still a truly terrifying figure when compared with the relatively mild case fatality rate of ~1 percent in the general population in 2020 for SARS-Cov-2, the RNA virus responsible for the COVID-19 pandemic.[12] If we pause for a moment and consider the huge disruption to our lives that the COVID-19 pandemic caused, it is almost unthinkable to consider what a more virulent virus might do to the world.

It is easy to forget that a 2014 Ebola outbreak in Guinea, in West Africa, quickly spiraled out of control, and eventually led to over eleven thousand deaths, including some in the United States and Europe.[13] Global travel networks now allow anyone to reach far corners of the globe in forty-eight hours, there is little to no medical oversight along the way, and the virus is essentially invisible in its early stages of infection. The initial symptoms of Ebola can take up to twenty-one days to manifest after exposure, and they are easy to overlook or dismiss. Ebola's clinical course is highly variable, but initially, someone may experience fatigue, weakness, muscle pain, chills, shakes, or a low-grade fever, and because viral loads in the blood are initially low, tests for Ebola may come back negative. But after a few more days, the symptoms can transmogrify into horrific fevers, maculopapular rashes (lesions resembling flat and raised red bumps), bloodshot eyes, severe abdominal pain and diarrhea, bloody stool, bleeding from orifices, and bloody vomit, which might require blood transfusions. During this ordeal, some patients might lose five *gallons* of body fluid, becoming dangerously dehydrated and contagious to unlucky helpers who come into contact with their bodily fluids, which are brimming with infectious viral particles. Eventually, blood vessels become so damaged that blood pressure drops, leading to hypovolemic shock, organ failure, and death.[14]

Because of the relatively inefficient person-to-person transmission via body fluids (compared with highly infectious airborne transmission like measles), a global pandemic from a future Ebola outbreak seems unlikely. But as the West African outbreak in 2014–2016 demonstrated, the wider world was not prepared, which led to panic and deaths.[15] Although several obstacles like access to the virus and its fragility outside of animal hosts make its use in a bioterrorism attack challenging, a determined and suicidal perpetrator could introduce the virus in a "closed population" like a cruise ship or health-care facility with devastating effects.[16]

The good news is that a vaccine was developed in the aftermath of the West African outbreak, and it seems to cut the death rate in half, even after infection.[17] I volunteered to receive it from the National Institute of Health (NIH) in February 2020, and I tolerated the shot just as well as any other vaccine. However, it will be logistically challenging and expensive to distribute it to the impoverished and rapidly growing African population that needs it the most. As relentless deforestation of Central Africa's forests continues, Africa's population swells, and climate change worsens, Ebola is likely to re-emerge from its hidden reservoir in a larger area of tropical Africa and more frequently—indeed, the number of outbreaks has already been increasing in recent years.[18] Only now do I realize how close our expedition came to being in the wrong place at the wrong time.

A few hours after we had left the leaping crocodiles of the Yenge River behind, we rejoined the larger Salonga River and stopped at a small village called Efofa, where a group of fishermen waved us down, shouting "*Nyoka!*" (snake!). In short order, we chased down two spectacular snakes, a juvenile Jameson's mamba (*Dendroaspis jamesoni*) and the aptly named splendid tree snake (*Rhamnophis aethiopissa*). The latter is a nonvenomous, arboreal species that reaches about 4.5 feet long; it spends most of its time well hidden in trees where it hunts birds, mammals, and lizards.[19] The splendid part of its common name is apt—this snake has a slim body with a contrasting pattern of lime-green, yellow, black, and aqua-blue flecks with large eyes.

Efofa was barely out of view when Poignard shrieked "*Nyoka!*" (snake!), to alert us that the young mamba had escaped and was crawling between our legs toward the stern! It was the only time during our journey together that an emotion of fear could be seen on his face. Wandege sprang into action and recaptured it within seconds, but I was not the only one who noted this was the second time he had let a highly venomous snake escape from his snake bag in the confined space of a boat. Chifundera scolded him mercilessly about checking his bags for holes in the future and I threw in my displeased agreement for good measure.

Our return to Watsi Kengo was welcome, and everyone reunited with friends we had made on the previous visit. I spent a productive evening recording forest treefrogs (*Leptopelis*) near our new campsite. When a massive photo shoot of the previous day's animals was completed the following afternoon, I had barely finished my lunch of river fish when a young woman belted out another shriek of "*Nyoka!*" (snake!) near our river encampment. Chifundera and Aristote tracked down the source of the cry and found a green bush viper (*Atheris squamigera*) perched on vegetation at about waist level above the forest floor. Reaching lengths of about 2.5 feet, these vipers have heavily keeled scales that are typically lime green with fine yellow crossbars. They lie in ambush in bushes and small trees for prey ranging from birds to other snakes. The venom is cytotoxic (capable of killing cells) and is known to be deadly to humans—two known fatalities occurred following six days of massive swelling and incoagulable blood.[20] Knowing the danger, we were especially careful as we photographed the irascible serpent, which attempted to bite us at every opportunity.

When I had a little downtime, I decided to use my Broadband Global Area Network (BGAN), a dinner-plate-sized device to connect to the internet in remote areas, to check my email. I noticed a message from UTEP president Diana Natalicio, who had just returned from a trip to San Francisco where she had a visit with my distant relative Richard Blum,[21] the wealthy husband of Senator Dianne Feinstein. Diana raved about his cherry-red Tesla, a novelty at the time, and passed along his comment that he wished I could "research frogs in a more easily accessible and safe place." Rather rich coming from a guy who fell in love with the Himalayas. Diana "assured him that wherever you were was likely more interesting than the places he'd wish you to be."[22] I could not have said it any better.

The collecting continued to yield many interesting things from our friendly hosts at Watsi Kengo, especially because the purpose of our visit had now spread far and wide, and we decided to stay for one more day. But on the morning of August 3, the fate of the expedition took an unexpected turn. I was enjoying my cup of coffee under a bright shard of sunlight that pierced the thick bamboo canopy of our campsite. But before my cup was empty, I could hear an increasingly loud and strained conversation in Swahili between Poignard and Aristote. Attracted by the trouble, Chif and Wandege emerged from their tents with concerned expressions.

"What happened?" I asked Chifundera.

He was so engrossed with the conversation and angry gestures of Poignard that my question took several moments to register. His mouth hung open slightly in a state of shock and I could see he was struggling to respond. His hand grasped his chin nervously as he finally said, "Someone is saying Poignard must be arrested."

"WHAT?!?" I replied. "What did he do?"

Another long pause, and then I could see his expression change to recognition of the situation, quickly followed by pity. He shook his head and said, "Ah, the women working with him set a trap, and maybe blinded by love, he was tricked."

The details of the plot remained murky, but apparently, Poignard had fallen for a get-rich-quick scheme, his judgment possibly clouded by duplicitous romantic overtures by one of the women. He had promised more than he could deliver, and now the women had reported him to the local authorities. Because they had home turf advantage, the women knew the police would side with them before outsiders like Poignard. And this is exactly what happened—they were judge, jury, and executioner, and they immediately determined that Poignard was guilty. They told him that he would be arrested unless he could immediately come up with the money to pay his debts to the women . . . and a sizable fine to the police. My sixth sense told me it was the work of the demanding woman in the red bandanna, and now I fully understood why she had creeped me out.

I could not help feeling pity for Poignard, and I understood his justifiable anger at the women for their manipulative deception. But within seconds, his seething rage boiled over in front of our eyes, and with clenched fists, searing eyes, and a contorted scowl, he hissed that he was ready to kill someone. All of us stared at him, flabbergasted, because it was clear that he was not kidding or exaggerating. Even the air around him seemed to shimmer in a maelstrom as we absorbed the real Poignard in all his terrible glory. Glancing at the expressions of everyone around me, I could see all of us were taken aback, and I feared what might happen next.

It was Aristote, perhaps the only friend he had in the world, who put his hand on Poignard's shoulder to calm him. I did not understand all the gentle Swahili emanating from Aristote's lips, but I could perceive him saying that Poignard would suffer a terrible fate if he really killed someone. Slowly, our cook's demeanor transitioned to defeat and resignation, but the vengeful countenance of his eyes remained.

In the end, I agreed to pay the debts and penalties for Poignard with an advance from his salary, he agreed to accept his mistake without violence, and we planned to leave immediately. After Aristote handed over the money to the police and women, we broke camp, loaded our pirogue, and hastened to travel west. The next time I heard anything about Watsi Kengo was in 2014, when the WHO declared an outbreak of Ebola in the area. A pregnant woman from a nearby village called Ikanamongo had butchered an animal from the forest, became ill, and died, but not before she passed the virus on to several other people. Eventually, cases spread to Watsi Kengo and Boende. By the time it was over in November, thanks to the rapid intervention of Congo's National Institute of Biomedical Research, the case count was limited to sixty-six with forty-nine total deaths, a very high fatality rate compared with previous outbreaks.[23]

Perceiving that he was on thin ice, Poignard was on his best behavior as we left Watsi Kengo behind. Hours into our journey, using a tiny campfire in the bow of our pirogue, he produced one of his most delicious meals ever. Misgivings or not, it was impossible to dismiss his culinary skills, and with no feasible replacement, I again saw no alternative to retaining him.

However, it was now crystal clear to everyone, as if there had ever been any doubt, that Poignard had deep-rooted anger from something in his past. Exactly what trauma had befallen him, or he had perpetuated on others, was still unclear. But everyone had witnessed the simmering volcano of his rage at Watsi Kengo, and if someone or something angered him again, an eruption of violence would surely follow. Firing and leaving him stranded in some remote village would be unethical and might trigger him, so I rejected this option, even though common sense demanded its consideration.

With my belly happy and full, I was still wrestling with the options to deal with Poignard when two men in a small pirogue approached us from the distant riverbank. Word had indeed spread about our search for snakes, and they presented us with a young, 3.5-foot-long Central African rock python (*Python sebae*) that they had captured in a net at the edge of the Salonga River. Rock pythons are Africa's largest snakes—two similar species occur in much of sub-Saharan Africa, but their exact boundaries, both geographically and genetically, are poorly known because they are CITES-listed species and rarely collected (see my previous comments about CITES permits and crocodiles in chapter 7). These snakes are known to reach over 24 feet in length, and some reports claim they can reach 32 feet. This is just shy of the record for longest snake in the world by Asia's reticulated python (*Malayopython reticulatus*), which can reach nearly 33 feet in length and prey on animals as large as sun bears.[24] Paul du Chaillu, an African American explorer in Gabon, said he killed snakes that were 20–25 feet long on "several occasions." He also reported a "boa" (undoubtedly a python) skin that measured 33 feet, but it is possible the skin had been stretched.[25] A British missionary named William Bentley noted that some pythons in Congo "sometimes attain a great length, even to more than 30 ft."[26] Mary Kingsley, another nineteenth-century explorer, was "assured by the missionaries in Calabar [Nigeria]" that they had seen a python that was over 40 feet long, but this record and other dubious ones above 30 feet have been criticized.[27]

Central African rock pythons are known to frequent areas with water and climb trees, but they are mostly terrestrial, and larger adults usually wait in ambush from a coiled position for prey ranging from a wide array of mammals to crocodiles.[28] Like other pythons, females lay eggs and will coil around them, both to protect them and raise their temperature during incubation.[29] Because adults can take relatively large antelope like Uganda kob (*Kobus kob*) and Thomson's gazelle (*Eudorcas thomsonii*), they actually have some overlap in their diet with lions.[30] Given their large size, it should not be surprising that they are consumed in many parts of Africa, and the fat is used as a treatment for rheumatism and earaches in South Sudan.[31] The English journalist Maurice Richardson likened python meat to "toughish veal."[32] The Dinka tribe, also from South Sudan, treat them as "privileged inmate[s] in the huts" and women will sometimes rub them with fat and even pour fat down their throats.[33] In other African cultures, pythons are revered, including Rwanda, where the souls of dead kings are said to endure in the majestic snakes.[34] We did not travel much farther before another fisherman approached us with another python, this one only slightly smaller than the first, suggesting either an incredible coincidence or that our visit coincided with the time of year when many of the eggs were hatching.

During one of the countless hours as we drifted past a stretch of forest with many raffia palms lining the riverbanks, something darted past my peripheral vision, snapping my vigilance back to the pirogue and our immediate surroundings on the river. Aristote was seated in front of me and as my eyes focused on his back, I saw an insect, about the size of a quarter, zip around his head with ninja-like agility. He instinctively swatted at the annoying bug, but it was incredibly fast and easily ducked away from his hand. And then it counterattacked by landing on his shoulder, and with a lightning-quick jab, bit his deltoid muscle.

"Ahhh!" he exclaimed as his shoulder twitched in pain and he flailed his hand to scare it away. "Shit, that hurt!"

"Aris, what is that?" I asked.

Before he could respond, more of the winged marauders showed up, and all of us were now busy in our attempts to shoo them away.

"Ahhh—they are tsetse flies!" opined Chifundera.

Terrorized by the prospect of sleeping sickness, and recalling Gloria's suffering, we redoubled our efforts to swat away the flies, but there were dozens of them, and every minute or two, one of us would wince in pain from a bite. Only Poignard managed to shield himself from the attack by wrapping himself up in his winter coat until only his hair was visible from the top. And then my luck ran out. Given the large needle-like proboscis of the flies, it is more accurate to say that one is stabbed by them, rather than bitten. The sharp pain emanating from an unreachable spot in the middle of my back was intense enough for me to rise out of my lawn chair, remove my hat, and wave it in every direction behind me. The pain subsided after a few minutes, but the hysteria lingered, even thirty minutes after the last fly disappeared as suddenly as the horde had arrived.

The Belgians had taken extraordinary measures to eradicate tsetse flies during the colonial era, sometimes with disastrous results.[35] Had we stumbled into a remote pocket of forest where the flies had managed to endure? And if so, had they passed along the worm-like, trypanosome parasite that causes sleeping sickness? This worry would haunt us for months, but I had managed to kill and photograph one of the green-eyed flies, and years later, an entomologist assured me that we had been attacked by a close relative of the tsetse in the Family Tabanidae, commonly known as horse flies. Although they are not the family responsible for sleeping sickness in humans, one of them is known to transmit a similar parasite that infects camels in North Africa.[36]

We spent the night at a tiny fishing village called Itafa, where we found several interesting frogs and an aberrant *Polemon* snake that was slowly crawling through the leaf litter near a sizable stream that drained into the Salonga. Unlike other *Polemon* we had found on the expedition, it had a dark brown body, it was over a foot long and more robust, and the collar transitioned from tan to cream from front to back. With the benefit of its DNA, my students and I later determined that it was closely related to *Polemon robustus*, the species found by a Batwa girl over 140 miles away at Lake Mai-Ndombe, but it was genetically distinct from it, strongly suggesting it is a new species.[37]

We broke camp shortly after dawn and continued west in good spirits. Eager to engage in conversation to pass the time, I spoke with Chifundera, Aristote, and Wandege about the animals we had found and those that we had expected to see but were inexplicably missing. One of the most common lizards in the mountainous border region of eastern Congo is the chameleon—several species are present, some more common than others, but it is rare for a day to go by without seeing one, even in areas that have suffered heavy deforestation near villages. But for some reason, chameleons (known as *kameleo* in Lingala and Swahili) seemed to be absent from the forests we had been searching for many weeks. Why?

Here and there when we inquired about the arboreal lizards, someone might have remarked that they had seen one the day before, or the week before, or even years before, but for reasons that eluded us, we had not been lucky enough to find even one. With plenty of time to think about the mystery, we came up with three hypotheses to explain our hard luck with chameleons. First, perhaps they are just rare, and one needs to spend huge amounts of time searching to find one. Before leaving on the expedition, I recalled looking over an opus on Central African chameleons by a Belgian herpetologist named Gaston-François de Witte, which showed an obvious scarcity of records from forests south of the Congo River, where we had been searching.[38] One of the few species known from the area was Owen's chameleon (*Trioceros oweni*), named in 1831 for a Canadian vice admiral who collected the first specimen from Bioko Island in Equatorial Guinea,[39] a species I had encountered only once before (with help from Mbuti hunter-gatherers) at Epulu in the Ituri rainforest of northeastern Congo in 2009. It is a relatively large chameleon (maximum length just over a foot long) that is sexually dimorphic—females are hornless, but males *usually* have three prominent horns projecting from their heads.[40] Like other chameleons in this genus, the horns are likely used for male-male combat when they are fighting over access to females.[41] Nearly eight decades after the description of *T. oweni*, a French herpetologist named François Mocquard received a specimen from Gabon with only one horn, and reckoning that it was distinct from *oweni*, he dubbed it *Chamaeleon unicornis*.[42] However, subsequent chameleon experts like de Witte decided the one-horned individual was just an aberrant specimen of the former species and considered the two names to be synonyms.[43] It seemed unlikely to me that the number of male horns should vary within a species of chameleon, but given the limited number of specimens available for comparison, there was some support for my hunch that chameleons are just rare in the lowlands of Central Africa.

A second hypothesis is that they might become active only during the rainy season, and aestivate (i.e., analogous to hibernation but in hot or dry weather) during the relatively hot and dry parts of the year. Several kinds of chameleons are known to do this, but most of them live in environments where the dry season is especially harsh.[44] In one extreme case of a chameleon (*Furcifer labordi*) from the arid southwest corner of Madagascar, the animals spend most of their lives (eight to nine months) developing in eggs during the dry season. When the rainy season arrives, all the eggs hatch, they reach sexual maturity in two months, quickly reproduce, and then die within another two to three months.[45] Perhaps we had not seen any chameleons because we were working in the dry season, and maybe climate change had exacerbated the conditions, encouraging the chameleons to aestivate longer. But given the lack of modern weather stations in Congo and adjacent areas of equatorial Central Africa, it is not clear if long-term downward trends in annual rainfall in other areas of Africa are having a major impact on the Congo Basin too.[46]

The third hypothesis is that the chameleons evaded detection because they spent most of their lives in the canopy, at least one hundred feet above the forest floor, where one would need a pair of binoculars to see them. Several kinds of chameleons do seem to spend most of their time in canopies of forests in Madagascar and we have already seen the example of seemingly rare, canopy-dwelling keel-bellied lizards (*Gastropholis*) in Congo. Then again, few searches of lizards in forest canopies have been conducted in Africa—such searches can be logistically challenging and dangerous—and some experts have argued that chameleon use of canopy habitats is based on anecdotal observations that are inherently speculative.[47] When we questioned a teenager who claimed to have seen a chameleon in one of the tiny villages we had passed through, he told us that it was climbing on a tree high above the forest floor and heading upward.

Perhaps a combination of factors from the three hypotheses was responsible for the dearth of chameleons on the expedition, but regardless of the reasons, it was disappointing because chameleons are spectacular lizards with fascinating life histories. Given our lack of time and data, all we could do was shrug and agree that it would be up to future workers to figure out which of the hypotheses, if any, were correct.

We passed by several familiar villages during our meandering return journey, including Bomputu, where we stopped only briefly to buy some bread and other supplies. I stayed in the boat, eager to move on, but as my eyes scanned the familiar people, huts, dogs, and other landmarks of the village, a new feature caught my eye. Clearly visible to all passing boats was a large wooden structure resembling a medieval palisade, complete with pointy sticks and a table-like structure made from branches and interwoven lianas. Strewn across the branches were the butchered bodies of several forest mammals ranging from monkeys to duikers. A wispy column of black smoke emerged from a fire at the base of the gory monument, which seemed to serve the dual purpose of smoking the meat and attempting to hold a frenzied cloud of black flies at bay. One of the animals seemed to be enormous compared with the others, and it had a distinctive reddish-brown coat with a dozen thin white stripes.

Aristote followed my stare and lamented, "Aahhh, very bad people are killing bongo for the bushmeat market."

"Bongo, the antelope?" I asked.

"Yes, the very big antelope that only lives deep in the forest," he replied sadly.

Bongo (*Tragelaphus eurycerus*) are one of Central Africa's largest and most beautiful forest antelopes, and contrary to Aristote's contention, they seem to prefer secondary forests and forest-savanna transition zones.[48] Males are larger than females and can reach about ten feet in length and weigh close to nine hundred pounds, about the size of a horse.[49] Both sexes don slightly spiraled horns that lean toward their back to prevent entanglement in lianas and other vegetation while running in the forest. They are capable of rearing up on their hind legs to reach vegetation that is up to eight feet above the forest floor. There are two subspecies of bongo, including the widespread *T. e. eurycerus* in the lowlands of West and Central Africa, and the slightly larger *T. e. isaaci* in shrinking and isolated montane populations of East Africa.[50] Despite enormous gaps in the geographic distribution of bongos in West, Central, and East Africa, nobody has compared the genetics of these populations to determine whether cryptic species might be present. A 2016 IUCN Red List assessment of bongos considered them to be "near-threatened" because of hunting pressure (both for food and trophies) and habitat loss, but given inadequate monitoring and unresolved taxonomy, their true level of extinction threat is probably underestimated.[51] It was truly heartbreaking to see a bongo reduced to a grisly offering in a riverside bushmeat market, and I could not suppress the terrible thought that this might be the only time I would see one in the "wild."

The title of a recent conservation study posed the question, Are we eating the world's megafauna (i.e., larger animals) to extinction?[52] Unfortunately, the answer is yes. On a global scale, only about 20 percent of tropical forests are considered to be "intact," but even fewer of these forests still contain healthy populations of wildlife. Of course, we have been hunting wild animals since the dawn of humanity, but in the twenty-first century, the hunting pressure has become unsustainable because there are too many of us compared with the animals, bushmeat is now commercialized, and there is unprecedented access to areas that used to be considered remote in previous centuries. The hunting pressure is most intense on larger mammals (heavier than 44 pounds), and parts of West and Central Africa have some of the highest defaunation rates in the world, with very few unaffected areas remaining in Congo. Removal of these larger species (including carnivores and herbivores) is likely to disrupt important ecological processes tied to the food chain and seed dispersal, which will almost certainly lead to negative domino effects to the composition and integrity of the forests, including extinctions of plants and animals. This is bad news for everyone, because damaged tropical forests (and the biodiversity they contain) store less carbon, which in turn worsens the climate crisis.[53]

We crossed from the Salonga to the Busira river as dusk approached, and stopped at a small fishing village called Bokengu to ask permission to pass the night. Delighted to have company, the village chief granted our request, and his three teenage daughters offered us *pombe*, creating a huge morale boost for everyone but me. I had started a course of antibiotics to get rid of an annoyingly persistent stomach infection, and I did not dare to jeopardize my recovery with alcohol. Exhausted and feeling unwell, I excused myself to set up my tent at the edge of the forest just as the grayish light of dusk yielded to darkness. In the dim light of my headlamp with fading batteries, I did not notice until it was too late that I had pitched my tent right next to a place where someone had recently defecated. Too exhausted to care, I retreated into my tent, removed my grimy clothes, rallied enough strength to inflate my air mattress, and passed out.

During the night I stirred once or twice during a raucous thunderstorm, but when I awoke in the morning, I was not prepared to find that my air mattress had become an island in the middle of a large puddle of rainwater that had leaked into my increasingly battered tent during the downpour. As I sat up to assess the flood damage, the weight of my body pushed the center of the slightly deflated air mattress lower, allowing the water to invade the depression in the island and soak my boxers. Cursing as I dipped my socks into the unavoidable mess and scrambled half-dressed out of the tent, I spent the next half hour trying to ignore a headache as I removed all my gear from the tent, tipped the water out, and moved it to another seemingly turdless place. By the time I was done, I was actually grateful that nobody had risen from their hangover-induced coma to witness my humiliating predicament, but it was small consolation for the increasingly difficult circumstances.

We had now been on expedition for nine weeks and all at once I started to notice how difficult things had become for myself and the team. Exhaustion and illness had sapped my vim. Looking around my tent, I could see that everything was covered in white mold, including most of my gear, the lining of the tent, and some of the precious snack food I had brought from the United States. Because Poignard did not have much experience washing clothes, a task usually assumed by women in Central Africa, my clothes had never really been clean or dry for weeks. As a result, the slimy biofilm on my skin had been challenging to combat, and with the benefit of a mirror and my headlamp, I discovered a new skin infection. Multiple infections are a warning sign that the body is weakened, and it was clear that I badly needed some R&R to recover.

And yet, we were still smack-dab in the middle of some of the most poorly explored lowland rainforest in Central Africa, and we had only a few more precious days remaining before we had to return to Mbandaka and arrange for a flight back to Kinshasa. If we dashed back to Mbandaka now, I could recover in relative comfort, but I would waste several days collecting in low-quality habitats around the city before heading back to Kinshasa. Instead, if I mustered the last of my strength and continued on just a little longer, perhaps I could find just one more spectacular species. I was sick and uncomfortable, but I was not yet on death's door and it was an easy decision to continue with the survey. Which dark green corner of the jungle would we search last?

After my African colleagues had a chance to drink some coffee and rejoin the land of the sober, I posed the question to Chifundera. He produced an ancient fold-out map from his backpack and pointed to the spot on the south bank of the Busira River where he thought we were. He slowly traced the path of the river to the west, until it bifurcated into a southeasterly branch called the Momboyo. He smiled at me as his finger stopped at large font text in the area just north of the Momboyo, which read "PYGMÉE." Recalling how friendly and helpful the Batwa had been to us in other areas of western Congo, I returned his smile and nodded.

A TWA'S GIFT FROM ABOVE

During the late morning of August 7, Jonathan steered our bow into yet another tiny fishing village with about a dozen people and a handful of goats and dogs. He did not explain the reason for the stop, but given the urgency with which he questioned one of the men before running off to the back of a hut, I guessed well enough. As we sat in our pirogue and waited, I noticed a thin woman in her late twenties looking at me more than the other people, and she raised her hand and addressed me in Lingala.

"Oh!" Aristote exclaimed.

Still unwell and a bit grumpy, I raised my hands in silence as if to say "Now what?"

Aristote took my gesture as a tacit signal for him to translate. He explained that the woman was offering to be my girlfriend—in a manner of speaking—in front of the entire village. I guess in Congo some women are not shy. For a moment my jaw hung open and I was too shocked to speak. Everyone in the village stopped what they were doing, seemingly intrigued by the offer, to stare at me in anticipation of my response. Even the goats and dogs eyed me impatiently. I glanced at Aristote and he was smiling, and Chifundera turned around with raised eyebrows to gauge my reaction.

In my diplomatic declination, I explained that the woman was very lovely and I appreciated the offer (especially while ill and looking like hell), but if I accepted, my significant other would first kill me and then move on to her.

172

And worse, maybe she would be so vengeful that she would kill all the goats, too, just out of spite.

When Aristote translated my response to Lingala, we all shared a good laugh, and poor Jonathan had to scratch his head in confusion when he returned to the pirogue. Even Poignard was laughing, the first and only time I saw him doing so. One little girl, however, took the threat so seriously that she grabbed her favorite goat, and crying with fear, led it away from the riverbank to safety. We all waved goodbye with big smiles as we pushed off into the river to continue our westward journey.

We stopped to resupply in the relatively large village of Ingende where the Busira and Momboyo rivers split, and for an hour I watched a group of barefoot women washing their dinner plates, dirty clothes, and children at the bank of the Ruki while I waited for Poignard to buy some food and other supplies. We motored southeast and up the Momboyo, but I quickly realized that my gamble would probably not be a good one. The forest had already been decimated on both sides of the river, and we could see many fires burning away more brush. When we reached the small town of Boteka, I could see why—remnants of colonial-era palm plantations (*Elaeis guineensis*) were everywhere. In general, palm oil plantations are bad news because the forest must be cleared to make way for the cash crops, which forces most of the animals that live there to either move away or die.[1]

As we continued east, I could see several rusting ships with vegetation sprouting from enormous cracks in their hulls piled alongside one another on the riverbank next to a broken crane on the side of a large dock, which was also in a state of permanent disrepair. Nearby in a grassy field at the edge of the river, I spotted huge metal tanks that were two stories high. They must have stored copious quantities of palm oil decades ago. Unbeknownst to me at the time, Boteka was the smallest of three plantations (the other two were Yaligimba and Lokutu near the northeastern arc of the Congo River) that were established in 1911 by William Lever, founder of Unilever, who received a land grant from the Belgian colonial authorities. At least some of the local villagers near these sites were forced to work on the plantations, and according to the Oakland Institute (a think tank in California), the people who live there now say the land was seized by the Belgians and should be returned.[2]

Despite the scene of decay, however, I saw a new truck parked near a renovated colonial-era house, which had the rare luxuries of power lines and a gleaming new satellite dish. Chifundera had learned at Ingende that a group of Canadians was now running the plantation at Boteka. In 2009, the plantations were being managed by Plantations et Huileries du Congo (PHC), and Unilever sold its stake in the PHC plantations to Canadian company Feronia after European banks provided the latter company with $150 million to finance the sale.[3] Much of the damage I had seen along the Momboyo had likely occurred long ago, but the new operation was clearly preventing the forest from regenerating.

Palm oil is essentially supercheap vegetable oil, and it is found in an array of products that are in high demand across the world, including cosmetics, shampoo, detergent, soap, toothpaste, infant formula, pizza dough, instant noodles, packaged bread, cookies, chocolate, ice cream, various kinds of junk food, and even biofuels. Some companies know all too well that it is unpopular to include an ingredient that is widely recognized to destroy rainforests, push endangered species closer to extinction, reduce water quality, and release greenhouse gases. Thus, they try to disguise it by calling it various other names like vegetable oil, stearate, sodium kernelate, and even the Latin name "Elaeis Guineensis," but usually, the name has some variant of the word "palm" in it, including palm fruit oil, palmate, sodium palm kernelate, and so on.[4] Even worse, some well-known companies (Unilever, Ferrero, Nestlé, PepsiCo, Weston Foods) associated with popular cereals, ice cream, and Girl Scout cookies are purposefully misleading consumers about "certified sustainable" palm oil, and sourcing the product from plantations suspected of employing child laborers.[5]

Palm oil plants originated in equatorial Africa, but in recent decades, industrial-scale plantations have been established in Southeast Asia, where the devastation has been the worst in the world. For example, in Malaysian Borneo between 1972 and 2015, new palm oil plantations accounted for 50 percent of the area's deforestation. Because palm oil can be grown in a wide array of soils where few other crops can flourish, it has replaced peatlands (recall the importance of this ecosystem from chapter 5) in many areas of tropical Asia, which has released a devastating amount of greenhouse gases into the atmosphere. Biodiversity in palm oil plantations is substantially lower than in natural forests, especially when the plantations are far away from intact forests, and many endangered species are rapidly losing suitable habitats, including orangutans (*Pongo* sp.), gibbons (*Hylobates* sp.), and tigers (*Panthera tigris*). A suite of negative environmental impacts accompanies palm oil plantations, including increased risk of fire (linked to a hotter and drier local climate from loss of tree canopy shade and other factors), pollution from fertilizer, pesticide, and other chemical use, and decreased hydrological networks and water quality. Although some Asian smallholders who grow palm oil see economic benefits, others must endure land grabs, violence, social inequality, and declines in quality of life.[6] Given the skyrocketing demand for palm oil, its inexpensive cost, and relatively high yield, more pristine rainforest is likely to give way to the crop in the future.[7] In 2024, a new analysis from the international NGO Global Witness and nonprofit organization Trase[8] concluded that palm oil imports to the United States are now the leading cause of deforestation in Indonesia.[9]

Like any commodity, palm oil has had periods of boom and bust, and only ten years after purchasing the plantations, the price of palm oil crashed and Feronia went bankrupt. In 2019, the banks stepped in to sell (at fire-sale prices) control of the PHC plantations to Kuramo Capital Management (KCM), led by Walé Adeosun, a former member of President Barack Obama's Advisory Council on Doing Business in Africa. In 2021, a KCM press release vaunted PHC leadership was transferred from foreign investors to "Congolese hands." According to a 2022 report from the Oakland Institute, KCM had powerful investors, including the Bill and Melinda Gates Foundation, and endowment funds from the University of Michigan, Northwestern University, and Washington University (St. Louis). The report also included allegations of physical abuse against local people, including murder, and at least one incident where PHC dumped toxic chemicals near a village.[10] In 2021, according to the Oakland Institute, KCM filed a lawsuit against one of its financial partners for "significant financial damages," and they responded with another lawsuit against KCM that is seeking $158 million in damages. Most recently, a lawsuit filed by a PHC shareholder accused Adeosun of money laundering and this person is seeking damages of $25 million.[11]

We traveled east as far as we could from Boteka before impending darkness forced us to find a place to spend the night, which turned out to be a Batwa village called Bolondo-Bulembe. The local people seemed very friendly, but they warned us that I would face jail and a huge fine if I tried to photograph them. Apparently, another white man had done so without their permission in the past, and things had not ended well for him. I told them I would respect their wishes, and with time, we gained their trust and they helped us find some animals.

The forest had looked ok from the river, but after we established our camp and had a chance to explore it more carefully, we realized it had probably been part of the original palm oil plantation, and was only beginning the initial stages of regeneration. Because of the sparse number of tall trees and abundant sunlight that still penetrated to the forest floor, thorny creepers and palms were abundant, and our clothes, boots, and belts snagged on the sharp thorns at every turn. At one point we lost our way and stumbled around for an hour before finding our way back to camp. My shirt was shredded by the thorns and my bleeding, scratched hands looked like I had been in a fight with a gang of cats. Aside from a white-lipped frog (*Hylarana albolabris*) that got spooked and tried to jump into Aristote's tent, we found nothing.[12]

The next day was a little better, but the pickings were slim. In the morning I was initially optimistic when I heard plenty of birds that frequent rainforests, including hornbills and turacos. A small group of Batwa women brought us a harmless, foot-long olive marsh snake (*Natriciteres olivacea*), a common species that they had found near a stream in the forest. We searched all day but found only three species of frogs, all of which we had encountered several times before along the Salonga River. The only lucky find involved green-eyed frogs (*Cryptothylax greshoffii*). For some reason, we had only encountered females the entire summer, but here we found a handful of males at the edge of the river.

I had to assert my authority when Wandege and Aristote disappeared to socialize with Poignard or drink beer instead of working. Chifundera seemed to be in need of rest because he was spending an increasing amount of time in his tiny one-man tent. I ate a bad bunch of bananas for lunch and suffered the consequences well into the night.

Amid this misery, Aristote came to check on me, and to pass along a concerning insight, yet again, into Poignard. He explained that while they were drinking beer, Poignard got tipsy and was bragging about marching into Kinshasa with Laurent Kabila's army to overthrow the government of Mobutu in 1997. This was consistent with his Rwandan accent because at that time during the First Congo War, Kabila had joined forces with the Alliance of Democratic Forces for the Liberation of Congo-Zaire (AFDL), a coalition of anti-Mobutu dissidents that was formed in Rwanda in 1996, shortly after the genocide in that country.[13] Rwandan Tutsi soldiers and the AFDL, eager for revenge against Hutu refugees, only some of which were responsible for the genocide (i.e., Interahamwe militia), slaughtered tens of thousands of them in a series of massacres across eastern and northern Congo, including the one I mentioned at Mbandaka.[14] When the AFDL entered victoriously into Kinshasa, they arrived with so-called *kadogo*, or Congolese child soldiers that they had trained to be hardened fighters. Part of the training involved Rwandan officers encouraging them to kill deserters who tried to escape the brutal conditions at the training camps.[15] Poignard had inadvertently admitted to being a part of this infamous group of genocidal fighters.

"We must be very careful with Poignard," Aristote emphasized, "because we do not know what he can do if he is very angry."

"So this is why he lied about his identity," I surmised. As the dots started to connect in my mind, I added, "Maybe he is a fugitive because he killed people during the war. This might explain why he snuck onto our boat at Kinshasa, why he disappeared when the DGM showed up in Mbandaka, and why he refuses to go to a clinic."

"Yes!" replied Aristote. "This must be why he was ready to kill someone at Watsi Kengo too! Maybe he did something *very* bad during the war, and now he is trying to hide it from us and people who *mayyyyybe* want revenge."

But who would be looking for him in this remote corner of Congo? Given the lack of justice we had seen during our journey, did Poignard really need to fear the long arm of the law? Maybe that is why he had left Kinshasa when we first ran into him. Then again, lacking concrete evidence, we might have been letting our pessimistic imaginations run away with unfounded conclusions. In the end we decided it would be best for all concerned to do nothing, but to stay very vigilant until we could end our work with Poignard. We were already very close to the end. Nevertheless, I had even more trouble sleeping as I thought about Poignard's past and the fact that he was preparing our food every day. Some of these fears would prove to be prophetic.

At dawn, I was so demoralized from my illnesses, disturbing revelations about Poignard, and bad luck for the final site of the expedition that I could not bring myself to unzip my tent when I heard someone approaching. I assumed it was Poignard with hot water for my coffee, so I was surprised to hear Chifundera's voice.

"Eli, can you guess what I have?" he asked.

"A black dragon roll of sushi and chocolate mousse for dessert?" I responded, knowing full well that he would not understand the nuances. My caffeineless brain was just starting to function, so it took a second to realize that the tone of my voice was truly pitiful. The combination of bad luck, fatigue from nine weeks of jungle fieldwork, three infections, and countless cuts and scratches from the thorns had demoralized me badly.

"What?" he responded, slightly confused by my American food references. "No! I have a chameleon!"

This time my brain reacted instantly, helped along by a surge of adrenaline, and in seconds I had forgotten all my troubles and was racing to throw on my moldy clothes. I was in such a hurry that I nearly forgot to slam my inverted boots together several times, a habit I had picked up in Bomputu to ensure that a scorpion had not crawled into them overnight. When I emerged from my tent, I was already breathing hard with anticipation, and I saw a smug Batwa sitting with Chifundera. Both of them stared at me, seemingly amused by my frazzled appearance. Perched on Chifundera's finger was a medium-sized green chameleon, and when my laser-focused attention centered on its head, I was shocked to see only a single horn protruding from it.

"It's *unicornis*!" I shrieked like a five-year-old on Christmas Day.

I stared at it in complete shock as the ramifications for Central African biogeography and chameleon taxonomy raced through my head. Time seemed to stop as I scrutinized every feature of the animal. It sure looked different from *Trioceros oweni*, the species that typically had three horns in males. Most of it was dark green, but parts of the head and limbs were lighter green, and a series of white spotty bands were present on the shoulder, along the body, and on the tail. The top of the head and the eye were grayish brown with rusty red speckles, some of which seemed to form bands too. I had never seen anything like it. I searched the head very carefully to see if there was evidence of broken horns or tiny ones, but it was clear that it had only one very prominent horn. It was substantially smaller than the *T. oweni* I had seen in Epulu years before, and the tail seemed to be unusually long.

My excitement roused Aristote and Wandege out of their tent, and joining the celebratory atmosphere, we stared at one of Congo's rarest chameleons with veneration. The Batwa man eyed us curiously, undoubtedly perplexed by our joy from a lizard. We asked him for the details about the capture, and he told us that he spotted the animal as it was crossing a footpath in the forest. He had found it about six miles away from Boteka in pristine forest, not a surprise given the shabby quality of the habitat where we were standing. I contemplated how news of an eccentric white man looking for reptiles had spread so far so quickly, but as I had seen before, word travels fast in Africa, even in areas where rapid communication seems to be impossible.

I danced a little jig to celebrate as I reached out my hand toward Chifundera for the creature to crawl on to my finger. When I took it from him, I immediately felt strange vibrations and a low humming buzz emanating from somewhere on the chameleon's body. I had experienced similar sensations after capturing dwarf chameleons (*Rhampholeon boulengeri*) in the mountains of eastern Congo,[16] but never before with a *Trioceros* chameleon like this. I suspect it is a last-ditch effort to frighten predators away, and because it is present in these two distinct lineages of chameleons, along with several others, the behavior must have evolved as a successful adaptation in the earliest origins of chameleons, which date back to the mid-Cretaceous when dinosaurs reigned supreme.[17]

Unbelievably, a fisherman brought us another chameleon a few hours later, and this time, it was a hornless female that had been found in a tree about ten feet off the ground. The forest gods certainly smiled on us that day, or more accurately, the forest people who helped us locate these rare animals. The benevolent feeling was reciprocated the next day when a small group of Batwa women from Bolondo-Bulembe passed by my outdoor laboratory and told Aristote how happy they were to have us there. When people from other villages had interacted with them in the past, they were often treated with disdain. But apparently, because we had come and opted to stay with them, they now garnered the respect of every village along the Momboyo. When they returned from a foray into the forest to gather edible mushrooms, they presented me with another *Helophis* water snake, which I accepted with sincere appreciation.

The day before we left, a passing fisherman stopped to let us photograph a rarely seen turtle that he had captured from hundreds of miles away in eastern Congo. Looking like the turtle equivalent of a huge (over a foot in diameter) black pancake with a fleshy and orangish head, I recognized it as Aubry's flapshell turtle (*Cycloderma aubryi*). This species was named for a well-traveled French civil servant named Charles Eugène Aubry-Lecomte in the mid-nineteenth century, and it is in the family Trionychidae, which includes the softshell turtles of the southeastern United States.[18] In general, these turtles have a flattened, circular body with a leathery shell that facilitates their highly aquatic lifestyle and powerful swimming abilities. Softshells are also known for their unusually long necks, fleshy lips surrounding the jaws, and a snorkel-like proboscis, sometimes used to snatch a breath of air while staying concealed underwater. Many of them can stay submerged underwater for long periods of time because of their ability to exchange gases via their skin, cloaca, and buccopharyngeal cavity.[19] Flapshells are a subgroup of African and Asian trionychid turtles in the subfamily Cyclanorbinae that are distinguished by the posterior edge of the underside of their shell (i.e., plastron), which have flaps made of connective tissue that are flexible, affording extra protection to their hind legs and tail.[20]

Owing to their penchant for spending most of their time in reedy coves with turbid water, along with the general lack of turtle biologists working in Central Africa, the African flapshell turtles are among the most poorly known trionychid turtles in the world.[21] Reaching a maximum size of about twenty-four inches and weighing up to forty pounds, Aubry's flapshell turtle has a large geographic distribution that is clustered around the Ogooué and Congo rivers in Central Africa, mostly in areas with tropical forests.[22] They feed primarily on fish and detritus, and like all turtles, they lay their eggs (seventeen to thirty-four per clutch) on land.[23]

The turtles are currently listed as vulnerable to extinction because of two main threats, including the bushmeat trade, and more recently, the pet trade.[24] A German scientist who conducted interviews with elderly fishermen in Gabon in the 1990s revealed that the turtle populations were already nosediving.[25] More recently, juvenile flapshells with colorful shells have started entering the US pet trade for $2,000 each. Wild adult turtles that have been labeled as captive-bred animals have been confiscated at New York's JFK airport.[26] Little research has been done on flapshells in Congo, but a turtle survey team documented over one hundred of them from the Luilaka River at Salonga, and half a dozen were seen at a bushmeat market in Goma in eastern Congo.[27] At the end of my expedition, I was shocked to see them in a fish tank, apparently on the menu, at a Chinese restaurant in Kinshasa.

The discouraging status of flapshells is similar to many of the other circa 365 species of turtles and tortoises (i.e., chelonians) around the world, which have a collective history dating back 220 million years, longer than mammals and much longer than birds.[28] In a landmark paper from 2020 titled "Turtles and Tortoises are in Trouble," fifty-one experts painted a bleak picture for the future of the animals, finding that over half of them are threatened with extinction.[29] It is common knowledge that chelonians live exceptionally long lives and for many species, sexual maturity is delayed but long-lasting—in 2023 a ninety-year-old, critically endangered radiated tortoise (*Astrochelys radiata*) named Mr. Pickles sired three babies at the Houston Zoo.[30] These life-history traits for longevity have been serious disadvantages in the most recent era of human dominance on planet Earth. When populations of a given turtle species decline, it can take decades for them to recover, often requiring the assistance of captive breeding programs. Three species have already been rescued from imminent extinction in this way, and one ambitious program is trying to reintroduce Bolson tortoises (*Gopherus flavomarginatus*), the largest and rarest tortoise in North America, to my home state of New Mexico, where they were wiped out in the Pleistocene.[31] However, captive breeding programs and concomitant reintroduction efforts can be expensive, logistically challenging, and ineffective for some species, especially those for which no wild habitat remains.[32]

Chelonians suffer from the same threats as many other vertebrates, including habitat loss, pollution, climate change, and overharvesting of their meat and eggs, but they are also targeted for the pet trade and the enormous Asian traditional medicine market. As species become increasingly rare, they get caught in a catch-22 whereby rarity increases their monetary value, fueling poaching of the last wild individuals. Some species of the rarest Asian box turtles (*Cuora*), for example, are so coveted in Asia that they can fetch tens of thousands of dollars. Unfortunately, one species of tortoise (Pinta giant tortoise, *Chelonoidis abingdonii*) and several other chelonian subspecies have already slipped into extinction in recent decades, and it seems inevitable that we will lose more this century, but the extent of the damage can be mitigated by the actions or inactions we take now.[33] Given the lack of resources, weak law enforcement, and corruption in many parts of the developing world where threatened turtle species occur, the challenges are daunting, but it is not too late.

I had mixed feelings on my last night in the forest on August 10. On the one hand, I was approaching physical exhaustion, Poignard was increasingly problematic, my money was running low, and I knew it was for the best that the expedition would end soon. But another part of me did not want the journey to end. I would miss the primeval scenery along the rivers, the peaceful tintinnabulation of raindrops falling on vegetation around me, and the earthy smell of the forest after a torrential downpour. But most of all I would miss the surprises. I could never predict what unique creature might be hiding under a fallen log, scurrying across a tree, or slithering near a forest stream, just waiting to be discovered. I had become thoroughly addicted to the excitement of unexpected discoveries like the chameleon, and withdrawal would be brutal. I could not shake the nagging feeling that there was so much more to discover, and given the suite of human-caused threats I had seen, very little time to do it.

But while I was preoccupied with Congo's fragile biodiversity, I was clueless about troubling circumstances happening in my camp. Wandege had an argument with Poignard about money he had loaned to him for his radio, and instead of promising to pay him back, our cook denied responsibility and opted to withhold food as retaliation. The private conversations he was having with Aristote were much worse, and my ignorance of them probably saved us all from catastrophe. Serious trouble was brewing, and it would not take long for it to burst out into the open.

10 TRUE COLORS OF THE DAGGER MAN

The afternoon of August 12 found us back on the sprawling lawn of the Belgian house at Mbandaka where we had started our journey east so many weeks before, but this time, we were not alone. Several European tourists were staying at the house, and they were led by a white man who had completely embraced the Congolese culture. Tanned from head to toe and sporting an enormous gut with unabashed pride, he strutted around with shorts and nothing else, a bottle of Primus beer in his hand, and a petite Congolese girlfriend at his side. I learned he was leading a boat tour along the Congo River, and I noticed the tourists had adopted some of his skimpy attire, including short-sleeved shirts, shorts, and sandals. They were friendly and un-bothered by our tents on the lawn, and given my rugged appearance and stories from the river, they probably regarded me as an attraction too.

It was nice to chat with some people in English, but the tour guide seemed irked when I asked them if they were taking medication for malaria. I could not hide my shock when they said no, especially since they were exposed to ravenous hordes of mosquitoes at dusk. The tour guide said that he had quinine medication with him, and that he would use it if any of the tourists became ill. He also mentioned that he had known Belgians who lived in Congo during the colonial era, and they told him that they came down with malaria three times a year like clockwork. Back then, however, they had well-equipped hospitals and pharmacies with well-trained staff to care for them if the malaria took a turn for the worse, at least in the major cities. If the tourists became ill in the middle of the Congo River, I shuddered to think what might happen to them. I wished them well and returned to our campsite to relax before dinner.

After Poignard obtained some meat from a nearby market, I gave him a packet with Jack Daniels barbeque marinade that I had been saving for a special occasion for months. He must have tasted it, because with wide eyes, he came to my tent and dangled the half-cooked meat in front of me.

"Verrrrrryyyy goot, boss!" he said, stoking both of our bellies to grumble. I had never seen him so excited, and he seemed almost giddy.

While I waited for him to finish cooking the meal, Aristote came to my tent and asked, "Hello, boss, how are you?" I knew from his tone that something was on his mind.

"I am very hungry, Aristote, but we have good meat, and I think it is almost ready. Are you ready to eat?"

"Yes, I am hungry, too, but not as hungry as Wandege," he said with a slow shake of his head.

"Why is Wandege so hungry?" I asked.

"Well, some days ago, he asked Poignard to give him the money that he had loaned to him for his radio. Poignard refused and said, 'Ok, Wandege, now you will not eat.' Until now, he has no food."

"Are you kidding me?" I asked with shock and anger. For weeks I had adopted the habit of taking my meals inside my tent to avoid flies and other insects from bothering me and swarming the food while I tried to eat, and as a result, I had not noticed that one member of the team had apparently gone without meals. I went to Wandege's tent and found him resting. When I questioned him, he angrily confirmed that Poignard had not paid him back or given him any food for three days. When I asked him why he had not said anything to me before, he shrugged his shoulders with a pouty expression and said that nobody cares about Wandege. I assured him it was not true and that I would get him a very good dinner.

I told them to wait and returned to my tent. When Poignard brought my food, I told him to follow me, and he watched as I gave it to Wandege. He looked sheepish as I instructed him to feed the rest of the team and then bring me my share. He followed my instructions, but his enthusiasm for the tasty food evaporated. I thoroughly enjoyed my meal, but I decided it would be the last one I would eat from Poignard.

In the morning I paid off Jonathan, Poignard, and the boat owner, and Chifundera was able to purchase tickets for us to fly back to Kinshasa. We all breathed a little easier when Poignard took his money and left, and thinking he was out of the picture for good, Aristote decided to drop a bombshell. He told us that just a few days before, as our work along the Momboyo River was drawing to a close, Poignard had asked Aristote to join him in a plot to poison us and abscond with all the valuable gear and money they could carry. Of course, Aristote refused, and he wisely kept the plot a secret until we were safely rid of Poignard. If he had told me there, I would have fired Poignard on the spot, and the reaction surely would have been violent.

Chifundera shook his head in disbelief and said, "This man is dangerous!"

"Yes!" said Aristote. "Really, he *must* be a former Rwandan soldier."

I absorbed the gravity of Aristote's admission in silence. We had been very lucky that the tourists were around on the last night Poignard cooked for us, because he would not have dared to try anything with so many witnesses. Nonetheless, we had dodged a very dangerous bullet, and I felt numb as I realized the truly evil nature of the man who had been part of our team for weeks, and how close we had come to catastrophe.

To save money for more beer, we decided to spend another day in tents at the house, and we managed to catch a handful of West African rainbow lizards (*Agama picticauda*) that we spotted along the brick walls nearby. When the tourists left on their boat tour, I was able to take a splash bath in a secluded corner near the house, and without a cook, we ate at restaurants for the rest of the day, indulging in cold sodas and beer. After receiving a tip about the best food in town, we headed to the Restaurant Les Delices de Trois Soeurs (Delights of the Three Sisters), where I was surprised to find a mid-forties white man in charge. With a big smile and a charming Belgian accent, he introduced himself as Eli, an incredible coincidence.

After we devoured several beers and grilled fish plates, he sat down with me to chat, and I learned he had been running the restaurant for some time. His children had come from Europe to visit him and have a look around, but otherwise, he seemed to be on his own. I struggled to understand why he had been drawn to Mbandaka to run a restaurant with delicious but modestly priced food. When he learned that I had been in the forest looking for snakes, his eyes grew wide for a moment as he recalled a disturbing story from someone he knew in the logging industry. Some time ago, a man had been in a very remote area east of Mbandaka. He took a break, wandered into the forest, and was never seen again.

"Leopard?" I wondered out loud.

"No, it waz a snake!" he insisted.

"How do they know?" I asked. For a moment, my mind drifted with skepticism to the impossibly large snake killed by Arnold Schwarzenegger in the movie *Conan the Barbarian*.

"Becuz dere was no leopard footprints, and pythons had been seen fwequentwy in zee area."

To this day I continue to be skeptical that this man disappeared from a giant snake attack, because in general, they are rare.[1] However, I cannot dismiss the possibility out of hand. One source noted that African pythons attack man more frequently than any other giant snake, and therefore, it was surprising that no fatalities had been recorded.[2] But this is not true—several fatal python attacks have been recorded, but separating trustworthy accounts from yarns is challenging.

The historical literature is full of tales about man-eating snakes in Africa, some more credible than others, and sensational stories like these are likely older than written history.[3] The oldest account is from 1705, which described a python that had been killed in Ghana with a human in its stomach.[4] Another one featured the explorer Henry Stanley from *Heroes of the Dark Continent* in which "Stanley was horrified to see a huge python uncoil itself from the body of one of the black boys of the expedition and glide off quickly into the jungle."[5] A full-page, engraved illustration in this book depicts an enormous "Boa

Constrictor" (i.e., python) attacking Stanley and a group of his men from a tree, which seems to be an exaggeration based on the story of a man named Abedi, who was part of the retinue for famous slaver Tippu Tib.[6] In the comments to a similar illustration of unknown origin from 1856, herpetologists noted that some large snakes occasionally hunt while perched in trees, and therefore, might attack humans from this vantage point.[7]

Another vivid account of a python attack, including a photo of the snake's impressive skin, comes from Reverend M. Hunter Reid from Ngangila in western Congo,

> The snake's skin when dry was 25 feet 2 inches long, and 2 feet 7 inches wide. It is now in the New York Museum. . . . Just as the natives entered the woods or jungle to start up more game that huge snake knocked one man to the ground, breaking one arm and several ribs. It then threw itself about him and reached for another man, whom it also got into its embrace before I could get to the spot . . . I shot it but once. The expansive ball used blew out one half of its brain, and its motions on the ground were a sight long to be remembered . . . the stomach of the snake contained not less than one peck [approximately 2 gallons] of brass, copper, and iron rings, such as the natives wear on the arms and legs.[8]

British missionary William Bentley remarked "the great pythons hide themselves in the thick jungles or woods near to swamps and streams, and sometimes, on the upper [Congo] river, catch women who are fetching firewood."[9] Arthur Loveridge, a prominent herpetologist from Harvard University, transcribed information given to him by people living on Ukerewe Island in Lake Victoria, who claimed that python attacks were frequent, and two people (a frail woman and a boy) had been killed.[10] Albert Schweitzer, the famed German physician and philosopher who spent years in Gabon, provided information about a credible attack on a child—his mother stabbed the snake until it retreated, but she wounded her son in the process, and he nearly bled to death before a doctor saved him.[11] Bill Pruitt, an American missionary in the Belgian Congo, documented an attack by a "big python" on a man who was returning from a hunt with a small animal in his hand. During the struggle, the python wrapped its coils around his body and engulfed his hand, leaving permanent scars from its numerous needle-sharp teeth. Supposedly, the man survived because he had a long thumbnail that penetrated the snake's brain during the struggle.[12] A Uganda newspaper provided few details in a story about a "Lango youth" from Uganda who was killed and eaten in 1951.[13] The most recent case is from 2013, when a miner in Zambia was attacked by a "giant python" that supposedly threw him to the ground, almost killed him, and hospitalized him for a month, but details of this attack are suspiciously scant.[14]

The snakes can be grisly scavengers too—one nineteenth-century traveler in Congo noticed a fifteen-foot python eating a corpse in a village that had been decimated by smallpox.[15] In some areas of Africa, men even slept with their legs in a "V" shape, presumably to avoid being swallowed while they slept![16] More recent and reputable sources have simultaneously cast doubt on some historical accounts while also discussing the plausibility of python attacks. Two well-regarded herpetologists noted that the collapsed shoulders of an adult man are roughly 1.3 feet wide, and pythons larger than sixteen feet would be capable of swallowing them.[17] Presumably, smaller forest people would be easier to attack and consume, especially if the snakes could deform the shoulders during ingestion.[18]

A morbidly fascinating study about Asian reticulated pythons (*Malayopython reticulatus*), only slightly longer than African rock pythons, opened my eyes to the possibility of giant snakes being man-eaters, especially for relatively small Indigenous peoples. The first author, Thomas Headland, started working with Agta Negritos[19] in 1962 in Luzon, Philippines, and this group of Indigenous people had many similarities to African forest people. Adult males averaged 97 pounds, they lived in small closely related groups in temporary shelters, foraged in primary rainforest, and hunted wild animals for meat. Of the 58 men in the group, 15 of them were attacked by pythons, and 11 had "substantial scars" on their bodies from bites. All the men survived by counterattacking with knives or shotguns, but 19 people knew someone who had been killed by a python. In one especially disturbing attack, a python entered a thatched shelter at dusk, killed two of three sibling children, and was swallowing one of them when the father returned and killed the giant serpent with a knife. The study also considered python attacks on rural Asians in other countries, and two of them included victims who were struck from trees.[20]

When all of the evidence is considered as a whole, it seems highly likely that Central African pythons attack and kill people in Africa, at least occasionally. Given the size limitations discussed above, relatively smaller humans are probably most at risk, including women and children. I never lost any sleep worrying about pythons during my time in Congo, because I knew attacks were exceptionally rare and my stocky frame and hearty appetite pushed my body's circumference beyond the range of a palatable prey item for all but the largest snakes. Nevertheless, our ancient primate ancestors might have endured python attacks on a scale rivaling those of the Agta,[21] because there are numerous python fossils known from East and Central Africa over the last five million years or so when bipedalism was evolving in hominins (species more closely related to humans than chimps).[22] Lucy, probably the most famous hominin fossil, represented a species (*Australopithecus afarensis*) that might be the ancestor of our genus (*Homo*).[23] She was only about 3.5 feet tall and 60 pounds, a relatively easy meal for an adult python.[24] Because the snakes are incredibly difficult to spot when they are lying motionless,[25] surprise attacks on lone, weaponless individuals[26] would have surely been successful, at least some of the time. Our nightmares surely pale in comparison with theirs.

Our spirits were high the next day when we transferred everything to the airport, and after checking our baggage, we sat around just outside the airport building with the other passengers while we waited for the plane to arrive from Kinshasa. Chifundera got a phone call and ran over to us to relay some gossip he heard from one of the men he had befriended near the old Belgian house. Apparently Poignard had showed up after we left and told Jonathan that he wanted the empty jerry cans that were left over from the long river journey. Each of them was worth about $5, and we had told Jonathan he could have them as a bonus to his pay. When Jonathan objected, Poignard told him they would fight for them. Jonathan must have agreed because he had a clear advantage in size and youth, but Poignard knocked him out with one right hook.

I was enjoying a book in the shade of a palm tree when Wandege touched my arm to get my attention. I looked up to see a black-eyed Jonathan pulling up to the edge of the crowd on a motorcycle and he had a passenger on the back—Poignard. He was donning his black sunglasses and, expressionless, he hopped off the motorcycle with only a slight hint of discomfort from his festering leg wound and turned around to stare at us.

Reeling from a shot of adrenaline, I kept my eyes on our former cook, but asked Wandege, "What the hell does he want?"

"I don't know," he replied, slowly shaking his head in disbelief.

It did not take long for Aristote and Chifudera to notice him, too, but he just stood there, arms crossed across his chest, observing us, saying nothing. When I asked Chifundera what we should do, he dismissed me with a wave of his hand and said, "He cannot do anything, there are too many people here, including police, and we are leaving soon."

I realized he was right, but for another hour, we sat around in an uncomfortable staring contest, each of us wondering what Poignard was thinking. If it was intimidation, he was definitely winning, but why? What was the point? Why would he spend any of his hard-earned money to come to the airport to stare at us? I briefly considered approaching a policeman to see what would happen if I pointed Poignard out to him, but in the end, I decided to do nothing and wait him out.

At last, the airport staff called for passengers to enter the building and proceed to the boarding gate, and with one final glance to ensure he was not rushing us with a machete, I looked at Poignard for the last time. He was stoic as he watched us leave. I did not allow myself to relax until I could see Mbandaka disappearing in the distance below us as we climbed into the clouds above the

Congo River. Even then, despite my exhaustion, I was so troubled I could not sleep. In a way it seemed impossible that Poignard was really and truly gone. I never saw nor heard of him again.

When we reached Kinshasa, I finally had an opportunity to look over the enormous collection of photos I had taken over the previous ten weeks and take stock of the successful expedition. I had been incredibly lucky to see an amazing diversity of rare and potentially new species, and I was looking forward to unlocking the secrets of the collections in my laboratory to truly understand this diversity. I already knew that ramifications from this work would be very important to ensure the species would be around in the future. The scientist and Pulitzer Prize–winning author Jared M. Diamond noted that "all decisions about conservation, wildlife regulations and creation of new national parks" are based on information from taxonomy, geographic distribution, and variation obtained from natural history specimens.[27] As one group of German zoologists put it, "Only species bearing a scientific name are entities recognized by society and politics for conservation."[28] I agree that assigning correct taxonomy to natural history specimens is the prerequisite to safeguarding any species, and I saw this play out when some of my taxonomic contributions were considered during the establishment of the Itombwe Nature Reserve in eastern Congo in 2016.[29]

Indeed, the publications that would eventually result from examinations of the expedition's specimens and genetic samples, including several new species descriptions, would improve global awareness and influence future conservation efforts for them. There would be other positive outcomes too—one of the colorful green snakes found at Npenda and Nkala would be named as a new species to honor Chifundera, *Philothamnus chifunderai*, in 2023.[30] Aristote and Wandege were coauthors on this paper, and it was the honor of a lifetime for Chifundera, who had made my entire research program in Congo possible.

There is still an enormous amount of work to do—about 86 percent of Earth's species have not been recognized and described yet—but unfortunately, it is becoming more difficult for a variety of reasons.[31] One of my PhD students managed to sequence DNA from the one-horned chameleon at the Momboyo River, and it was genetically distinct from the *Trioceros oweni* chameleon I sampled at Epulu, suggesting the two lizards are not the same species. This is a very exciting and important result, but now the question is whether to elevate *T. unicornis* to full species status or name a completely new species for the one-horned chameleon. To figure out what the correct course of taxonomic action is, I need to compare the Momboyo and Epulu chameleons with the *T. unicornis* chameleon from Gabon, ideally with genetic data. Unfortunately, the original specimen from Gabon was collected many decades before genetic samples were included in standard procedures for expeditions in the 1990s. It can be very challenging or impossible to obtain DNA data from older specimens that were preserved in formalin if the museum in question is even willing to allow some destructive sampling for this purpose.[32]

Because CITES permits for chameleons add another layer of complexity and expense to the stressful burden of export permits for natural history specimens, some herpetologists do not bother to collect them.[33] When I searched VertNet, the go-to database for genetic samples and specimens in museums, there were no hits for genetic tissues of *Trioceros* chameleons of any kind from Gabon or neighboring Republic of the Congo. Of course, it is important to protect chameleons from overharvesting for the pet trade, one of the primary goals of CITES, but it is a shame that scientific research is affected by these well-meaning regulations. Someday when more samples are available for comparison, if my lab's preliminary data on these chameleons prove to be correct and I can link them to their own scientific name, they will undoubtedly require updated conservation efforts to protect them.

But I might be waiting a very long time before someone makes the effort to collect new genetic samples of chameleons in other areas of Central Africa. Expeditions like the one described in this book are becoming increasingly rare in many parts of the world, including Congo, and for several groups of vertebrates, new collection efforts are lower than they were during World War II. As these efforts wane, problematic gaps in the long-standing record of natural systems are created.[34]

One of the biggest challenges is funding. Like everything else, expeditions have become more expensive in recent years, and grant opportunities to fund them are becoming increasingly scarce and more competitive. Many developing, biodiverse-rich countries have a double whammy of scant resources and poor training opportunities for biologists. In the United States, respected government institutions, universities, and even museums are turning away from whole-organismal biology and taxonomy in favor of molecular fields with more funding opportunities and lucrative "indirect costs" from federal grants that increase institutional coffers. The US National Science Foundation now has only one highly restrictive funding path that is earmarked specifically for taxonomy, a standpoint that is completely at odds with the extinction crisis happening all over the world. As a consequence, centuries-old institutions that are struggling to document biodiversity loss are themselves facing extinction. In 2024, Duke University made the gormless decision to get rid of its state-of-the-art herbarium collections (825,000 specimens), citing high estimated costs ($25 million) to upgrade facilities and endow curators and staff.[35] But even with a 1 percent loss of its investments in 2023, Duke had an endowment exceeding $11 billion.[36] In response, one researcher from the Smithsonian asked, "What's next, the library?"[37]

Another issue is "biodiversity nationalism" whereby certain countries pass legislation that make biodiversity surveys extremely difficult or impossible.[38] Some countries view scientific research as a way to make money and they try to charge thousands of dollars for permits. Unable to use modest grant funding to pay for this and unwilling to pay from their own pockets, many scientists, including me, have no choice but to avoid working in these countries. The result is that huge portions of their biodiversity remain unstudied, scientifically name-less, and unprotected, and students in these countries who could benefit from professional mentorship and international collaborations lose opportunities to work on the poorly known species in their own backyards. Relying on outdated lists of species, or ones produced by poorly trained biologists, has been likened to "knowing the number of pills in a jar, but failing to know which are stimulants, depressants, painkillers or hallucinogens."[39]

Paradoxically, this is happening just as an increasing number of scientists are turning to collections to answer important questions related to evolution, environmental change, ecosystem function, and zoonotic diseases, all of which are highly relevant areas of research in the twenty-first century.[40] Many questions we cannot fathom now, or are impossible to answer with current technology, will be possible to address in the future, including detailed documentation of biodiversity loss as humans relentlessly destroy the natural world.[41]

Cutting-edge museomics approaches are shedding new light on very old specimens. The only known specimen of the world's largest gecko species, *Gigarcanum delcourti*, was collected around the 1830s from an unknown location. The estimated date of collection was tied to its unusual preservation as a dried and mounted specimen. DNA isolated from its femur bone allowed scientists to understand its evolutionary relationships to other geckos, including its taxonomic status as a unique genus, and identified its likely origin from New Caledonia. No recent trace of the animal has been found, and it is almost certainly extinct. If somebody had not bothered to collect this specimen so long ago, we would not know it had ever existed.[42] Moreover, museomics could not be conceived as a future technology in the nineteenth century, many decades before DNA was identified as the molecule of heredity.[43]

As I explained in my 2017 book *Emerald Labyrinth*, collections are very likely to provide a future opportunity to clone species that have gone extinct in the wild, and assuming their habitats still exist (or can be restored), it might be possible to reintroduce them. But to do this, we must first know that they exist, where they occur, and other aspects of their ecology and natural history that can be obtained only from natural history specimens.[44] This de-extinction might sound like science fiction, but we have already discussed ongoing efforts to do this with the woolly mammoth (see chapter 1), and RNA was recently isolated from a 130-year-old specimen of the extinct thylacine (a.k.a., Tasmanian tiger, *Thylacinus cynocephalus*), which could aid efforts to bring it back to life.[45]

Throughout this book, you surely noticed pervasive evidence of the human-induced pressures that are driving many species to oblivion, and most scientists now agree that we are the cause of the unfolding sixth major mass extinction event on planet Earth.[46] One jaw-dropping estimate suggested about 50 percent "of the number of animal individuals that once shared Earth with us are already gone, as are billions of populations."[47] In a study from 2023, Paul Ehrlich, a respected emeritus professor at Stanford University, said (along with first author Paul Ceballos) that we are experiencing a "mutilation of the tree of life," which is "destroying the conditions that make human life possible."[48]

The "mutilation" is especially egregious for amphibians, far more than other major groups of vertebrates. Based on consultations with over one thousand experts who provided data from their fieldwork experience and associated collections data, including me, an updated Global Amphibian Assessment (GAA) was published in 2023 and the results were truly shocking. Of the 8,011 species assessed, 41 percent were found to be threatened with extinction, and over 200 species might already be extinct. Another 11 percent of amphibians were "data-deficient," meaning there was insufficient data to determine their extinction risk, but many of these species are likely vulnerable to extinction too.[49] Habitat loss was the greatest threat to amphibians, followed by climate change and chytridiomycosis (a fungal disease responsible for population declines and extinctions), which sharply increased in prevalence in Africa at the turn of the twenty-first-century.[50] With new species of amphibians being discovered all the time, either through expeditions or studies of existing natural history specimens, there will soon be 9,000 described species of amphibians, and scientists are already busy assessing their conservation threats for the next GAA.[51]

Similar conservation assessments for reptiles lag far behind other major groups of vertebrates. In 2022, the first Global Reptile Assessment (GRA) by the IUCN Red List was completed, and the results showed that at least 21 percent of the 10,196 assessed species were threatened with extinction. However, there are now over 12,000 species of reptiles (circa 240 new species described per year in the last five years), leaving a large portion of them unevaluated, and a more substantial portion have outdated evaluations that might underestimate their current threats. One study found that recently described reptiles are more likely to be classified as threatened or data-deficient because they often have small geographic distributions. Nearly 1,500 species of reptiles were classified as data-deficient in the GRA, and similar to amphibians, many of these species are probably threatened with extinction, if they are not already extinct. Major threats to reptiles include habitat loss, environmental degradation and pollution, poaching for food or the pet trade, and the existential threat to all life on the planet, including us—climate change.[52]

On April 22, 2021, as the world celebrated Earth Day, the Biden administration announced ambitious goals to reduce greenhouse gas emissions by half of 2005 levels *by the end of this decade*.[53] The goal of this action was to limit global warming to 1.5 degrees Celsius or less (a goal of the 2015 Paris Agreement[54]), because above this limit, multiple scientific models agree the planet's ecosystem and climate systems will begin to unravel, deadly tropical diseases will become endemic in the Northern Hemisphere, and droughts and sea level rise will displace or starve at least 100 million people by the 2050s.[55] Many of us have probably forgotten (or never knew) that there were a million annual cases of malaria in the United States in 1937, and a changing climate could create favorable conditions for the deadly disease to return.[56]

All of this is bad enough, but have you ever paused to consider the potential *direct* effects of elevated greenhouse gases on your body? Throughout most of our roughly million-year history on planet Earth, humans have existed in a world with carbon dioxide levels that fluctuated between about 135–280 parts per million (ppm) in the atmosphere. Today we are above 420 ppm and increasing at an accelerated rate each year, exposing us to unprecedented carbon dioxide levels—at least since the Miocene over ten million years ago—which our bodies are ill-equipped to handle in modern environments.[57] Multiple studies have found that these ever-increasing concentrations of carbon dioxide, especially when they get worse in crowded urban environments, have worrying links to direct negative effects on our health. These effects include inflammation, cognitive problems, serious malfunctions of the proteome (the entire suite of proteins made by our cells), bone demineralization, calcification of kidneys, oxidative stress, sleepiness, anxiety, and endothelial dysfunction, a type of coronary artery disease that can cause chest pain.[58] More studies are needed to improve understanding of these links, but some of these effects might already be happening to us, including to sedentary children who are exposed to higher levels of carbon dioxide in poorly ventilated and crowded classrooms.[59]

As chronicled in Bill Gates's book *How to Avoid a Climate Disaster*, the world must quickly reduce greenhouse gas emissions to zero, but some of the innovative solutions to do this do not yet exist.[60] We have less time than we think—in 2023 the global average temperature exceeded the 1.5 degrees Celsius boundary of the Paris Agreement for the first time.[61] The boundary was exceeded again in 2024, which was declared the hottest year in human history.[62] Just weeks into 2025, new studies suggested the 1.5 degrees Celsius warming threshold was likely breached for good, and that "the impacts that scientists told us will happen around 1.5°C [of warming] are going to materialise."[63] These negative impacts will include an increased pace of extinction for vulnerable species of plants and animals, especially amphibians and biodiversity that is endemic to relatively cool, montane regions. As I noted earlier in this chapter, amphibians already face a number of threats that are worsened by climate change. Rising temperatures will push all biodiversity that occurs in montane areas onto "an escalator to extinction" that dooms them to move to higher and cooler elevations with suitable conditions until they approach the summit and there is nowhere left to go.[64]

Although scientists have done a remarkable job of raising awareness about the need to reduce greenhouse gas emissions to avoid the worst effects of climate change, many would argue that they have completely failed to educate the public about the equally essential need to reverse damage to natural areas and conserve remaining biodiversity, especially in parts of the world that are far removed from their own backyards. Earth's ecosystems and global climate stability rely on intact natural areas, home to millions of species of plants, animals, and single-celled organisms that maintain the fragile balance of all life, including declining populations of insects that are responsible for pollination of 80 percent of the world's flowering plants and about three-quarters of all crop species.[65] Tropical forests contain about two-thirds of the world's biodiversity, and collectively, they contribute so-called biophysical effects that can have a cooling effect on the global climate, including creation of clouds, humidity, and chemicals including aromatic terpenes that affect the atmosphere.[66]

Many of us are probably aware that trees sequester carbon dioxide from the atmosphere via photosynthesis and store it in their wood, branches, leaves, and roots.[67] But perhaps you did not know that about 75 percent of terrestrial carbon (more than the atmosphere and plant biomass combined) is stored underground in soil, and mycorrhizal fungi facilitate this transfer via their ancient (circa 400 million years) symbiotic association with plant roots. The establishment of this plant-fungal symbiotic relationship likely facilitated colonization of land by early plants and is associated with a tenfold decrease in carbon dioxide in the atmosphere starting in the Paleozoic Era. A recent study found that on an annual basis, terrestrial plants transfer 13.12 $GtCO_2e$ (gigatons of carbon dioxide equivalent) to mycorrhizal fungi, which was over a third of human-caused carbon dioxide emissions from fossil fuels in 2021.[68] In tropical rainforests, high plant diversity is associated with diverse fungal assemblages, but when the forests are destroyed to make way for crops like palm oil, the soil fungal communities are reshuffled, symbiotic fungal species decrease, and all of this affects the health and functioning of the entire ecosystem.[69] Ecosystem collapse is game over for most of its biodiversity, and it also threatens food and water supplies that humans need to survive.[70]

Although the tropical forests of the Congo Basin (over 1.1 million square miles) are much less than those in the Amazon (over 1.9 million square miles), they actually pack more punch to fight climate change. A landmark study from 2021 in the journal *Nature Climate Change* showed that the net carbon sink of the Congo Basin forests was *six times stronger* than that of the Amazon, because gross removals of carbon were nearly identical, but gross emissions in the Congo forests were half those of the Amazon.[71] This finding is consistent with another study from *Nature* that found higher carbon emissions in areas of the eastern Amazon that have been affected by deforestation and climate change the most, resulting in ecosystem stress and increased fires.[72] This comparison underscores the need to halt deforestation in *all* tropical forests, which store about one-fourth of the planet's terrestrial carbon.[73] In Congo, more than 90 percent of the total energy used from local people comes from firewood and charcoal, and pressure on forests is projected to increase in the future.[74] Because African forest tree species are better adapted to heat and drought than other tropical forests in the Amazon and Asia, they will likely fare better as climate change grows worse.[75]

For these forests to do their job of hoovering out the greenhouse gases we are belching into the atmosphere with increasing disregard, they need to have healthy levels of biodiversity to function. Many plants make large fruits to disperse their seeds and they require large animals to eat and disperse them, but as we have already seen, megafauna have been decimated across the world (see chapter 8), and some species of trees will likely go extinct as their former seed dispersers disappear. In general, larger seeds are associated with higher levels of wood density and carbon storage, but if they vanish, their replacements are rapidly growing, small-seeded plants with less carbon-storage capabilities.[76] This is just one of many examples of an "extinction cascade," in which the extinction of one species has a domino effect that causes the loss of other species. Together with other stressors like climate change, several extinction cascades can saturate an ecosystem with negative consequences until it no longer functions properly.[77] Most worrying of all, as biodiversity loss continues across the planet, it "could be great enough to threaten the biosphere itself, which is logical given species are the elements that give the biosphere its structure and function."[78]

Can regeneration of recently damaged tropical forests be a nature-based solution to solve these gargantuan problems? Not alone, but a tragic example from history demonstrates its extensive potential. As explained in the book *Jungle*, when European explorers like Christopher Columbus made landfall in the New World, they had insidious company, including measles, smallpox, influenza, and even bubonic plague. This globalization of disease continued with the arrival of African slaves, who brought typhus, cholera, diphtheria, and malaria. Within a century or so, nearly 55 million people, or about 90 percent of the estimated Indigenous population of the Americas, were wiped out from several waves of epidemics, warfare, enslavement, and famine resulting from societal collapse in an event called the "Great Dying."[79] As a consequence of these human reductions, including from many tropical landscapes, over 216,000 square miles were abandoned, fire activity decreased, and forests regenerated naturally, a process known as secondary succession. During this time, atmospheric carbon dioxide levels decreased by an estimated 7–10 parts per million and global temperatures declined by about 0.15 degrees Celsius—much of this global cooling trend has been attributed to the secondary succession.[80] This drop in temperature might seem modest, but it was enough to precipitate the "Little Ice Age" (circa 1600–1650 CE) when crops across Europe failed from the freezing cold.[81] Some scientists have pointed to this series of events as the beginning of the Anthropocene, a proposed geological epoch to demarcate the point in Earth's history when humans started imposing quantifiable "biophysical impacts on the planet."[82]

Carbon capture by photosynthesis of trees is likely to be one of the most cost-effective measures we can take to limit carbon dioxide concentrations across the world, and there is enormous potential to do this, including in Congo.[83] One study estimated about seventeen million square miles of canopy-cover trees can be supported worldwide under current climate conditions, which is over six million square miles more land covered by them now—that is double the size of Australia.[84] But for this to be effective, the most ideal species of trees have to be planted in the right place, and depending on additional goals (e.g., supporting local biodiversity), for the right reasons. In 2021, a French oil and gas company called Total Energies proposed to plant forty thousand trees in Republic of the Congo. The problem is that they are Australian acacia trees that might harm local biodiversity and water supplies, and actually increase local temperatures because they could absorb heat from grasslands.[85] Allowing forests with native species to regenerate naturally is often the best and least expensive solution, especially in the tropics where plentiful sunshine and rainfall nurture high growth rates.[86] Some of the most promising areas for this to happen are in Congo and other areas of Central Africa.[87] A 2024 study in the journal *Nature* estimated that, in certain areas of the world, an area of tropical forest larger than the size of Mexico could undergo natural regeneration, but only if we work together and allow it to happen.[88]

A 2022 report from the Center for Global Development, a think tank in Washington, DC, estimated that the value of the Congo Basin forests' removal of carbon from the atmosphere is $55 billion per year (this is likely an underestimate), which is about five times the annual government budget of Congo. Because Congo is not being paid anywhere near a fair price for these invaluable services to the global community, it should not be surprising that the government is pursuing economic opportunities like oil and logging concessions that will destroy some of these forests.[89] Given the seriousness of the problems of climate change and biodiversity loss, and the eye-popping costs (both in trillions of dollars and human lives) that we will incur if we do nothing,[90] I believe it is high time that the world starts to pay for the invaluable ecosystem services that Congo's forests provide to all of us. Because of governance challenges in Congo, it might be tricky to set up a system to do this, especially one that shares benefits with the poorest people who need it the most.[91] But carbon-offset programs that pay landholders to plant trees and improve forest management in tropical areas throughout the world are showing incredible promise and could be scaled up to have a truly global impact.[92] There are many other cost-effective options, too, including inexpensive stoves that reduce reliance on firewood from forests and improve human health.[93]

The people and natural world of Africa face many challenges in the coming years. Recently, several Francophone countries experienced coups. Russian mercenaries of the Wagner Group terrorized people and fought shadow wars in several countries (including Congo), Rwandan-backed M23 rebels seized Goma and Bukavu and threatened a wider war in eastern Congo,[94] wooden furniture demand in the United States drove Chinese logging companies to increase exploitation of forests in the Congo Basin, and Congo considered auctioning off thirty oil and gas blocks (including at least twelve in protected areas) to the highest bidder. All of this is happening as the continent's population is exploding—by 2050, one in four people on our planet will be African.[95] It will be incredibly challenging, and maybe impossible, to feed all these people while simultaneously conserving the forests and biodiversity that are required to stave off the worst effects of climate change.[96] It seems inevitable, unfortunately, that some of the amazing places and species I discussed in this book will disappear from the wild long before the end of the century.

And yet I am an optimist. I must be because, in the middle of writing this book, I became a father with the hopeful outlook that the planet will be a good place for my son to thrive long into the future. For this to happen, I believe some irreversible catastrophe will need to shock the world into action, because at the moment, most of us are too distracted to heed warning signs of the climate change and biodiversity extermination happening all around us. Once global attention is laser-focused on halting climate change above all other priorities, a crucial prerequisite to address the problem, the forests and biodiversity of Congo will have a key role to play in saving us from ourselves.[97] The sooner we get started the better because, globally, we are currently losing the equivalent of a soccer field of tropical forest *every five seconds* along with nearly all the biodiversity they contain.[98] In 2023, the largest losses occurred in Brazil (43 percent), Bolivia (9 percent), and Congo (13 percent).[99] Will our descendants forgive us if we fail to act while we have the chance?

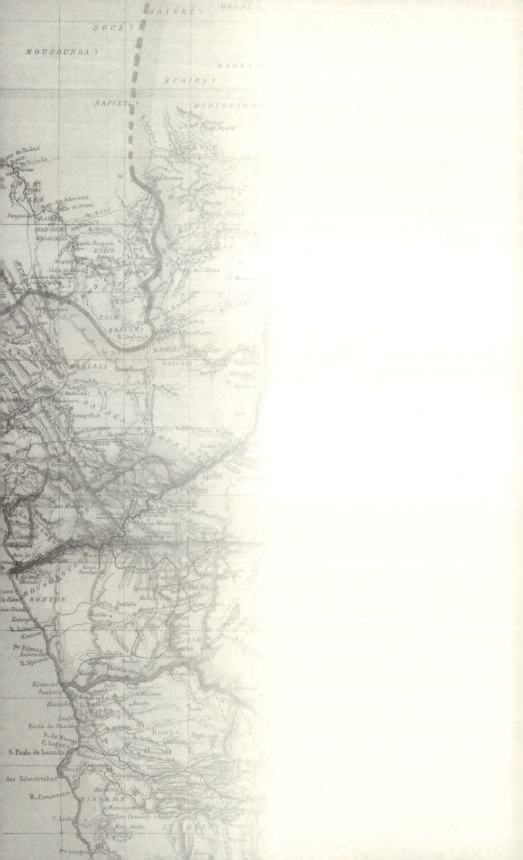

ACKNOWLEDGMENTS

My work in Congo would have been impossible without the collaboration, support, and friendship of my Congolese colleagues who are the main characters of this book, including Chifundera, Aristote, and Wandege. The Institut Congolais pour la Conservation de la Nature (ICCN) issued permits for the research. In 2012, the year before I commenced the expedition, I was awarded a major grant (DEB-1145459 with colleague Kate Jackson) from the US National Science Foundation to study the amphibians and reptiles of the Congo Basin, and many of the results I present in this book are linked to that grant. Because my students, collaborators, and I have now had over a decade to study at least a portion of the natural history specimens that were collected on this highly successful trip, many studies that are based on this material have been published. The data in these publications provide crucial updates regarding their identification, evolutionary relationships, and natural history that I did not know in the middle of the jungle in 2013 when the animals were first encountered. I am very grateful to the University of Texas at El Paso (UTEP) for supporting my seventeen-year endeavor to expand their natural history collections with my expeditions to Africa, which I sincerely hope they will treasure and protect long after I am gone. I appreciate UTEP Provost John Wiebe's faculty development leave program, which afforded me extra time to conduct research for this book in 2022. Former UTEP Department of Biological Sciences Chair Bruce Cushing supported my efforts to write this book while navigating new fatherhood and his successor Kendal Hirschi encouraged me as well. In 2022, I also secured a $500,000 University of Texas System Rising STARs Award, which funded some of the research that is discussed herein. Special thanks to UTEP College of Science Dean Robert Kirken (recently retired) for his support of the STARs award, and to Berenice Herrera, Jocelyn Gonzalez, and Hilda Orozco for managing the funds. Ana Betancourt of the Border Biomedical Research Center (BBRC) Genomics Analysis Core Facility helped my lab sequence DNA from hundreds of samples that were collected during the 2013 expedition. Special thanks to my friends and family who cheered me on during the expedition while harboring valid concerns for my safety.

As I worked on this book, my wife Denise supported my endeavors in countless ways, even when it was stressful to do so after the arrival of our son Damian, to whom this book is dedicated. The graduate students in my lab were understanding and patient in the final hectic months of writing, including Calum Devaney, Everett Madsen, Jesus Reyes, Dominic Troiani, and Tyler Blake. Undergraduate student researcher Angelica Casas assisted with research related to python attacks and helped to sequence many of the samples. Many other Greenbaum Lab students assisted with genetic data collection over the years, and I am grateful to them all. The book benefited immensely from readings and discussions I had with students in my graduate-level Biodiversity course at UTEP (BIOL 6312) from 2010 to 2025. Parts of the book were written during my eleventh expedition to Congo in 2023–2024; I appreciate the feedback and discussions I had with Chifundera and Aristote about our recollections of the 2013 expedition. Several people helped me to identify plants and animals from specimens and photographs, including, Lorenzo Prendini, Roy Gereau, Andy Plumptre, Piet Stoffelen, Jens Leifeld, and Brian Weiss. Several other colleagues provided crucial information or reviewed parts of the book, including John L. Carr, Lauren Esposito, Igor Almeida, Richard Wrangham, Colin Tilbury, Colince Kamdem, Carl Lieb, Scott Weinstein, Václav Gvoždík, Wolfgang Wüster, Zoltán Nagy, Tomáš Mazuch, Jo Thompson, Thomas Butynski, Erik Verheyen, Guy-Crispin Gembu Tungaluna, Tyler Jacobson, Michael T. Hernke, Douglas Watts, Terrence Demos, Harry Greene, Océane Da Cunha, Jean Claude Bimwala, Matthew Shirley, Steve Spawls, N. Thomas Håkansson, and Germán Rosas-Acosta. Thomas Frogh from International Mapping made the excellent map of western Congo. Adrian Soldati and Christian Ziegler allowed me to include their outstanding great ape photos. Nanna Heitmann allowed me to use her spectacular photo that first appeared in a *New York Times* story about Congo's peatlands (Maclean and Kabanda, 2022) for the cover of this book. Encouragement came at a crucial time from Marty Crump, Whit Gibbons, Arshad Khan, and my agent, Jill Marsal. Special thanks to Stephen Hull at the University of New Mexico Press, who chose to work with me again on this book after our mutually positive experience with *Emerald Labyrinth*.

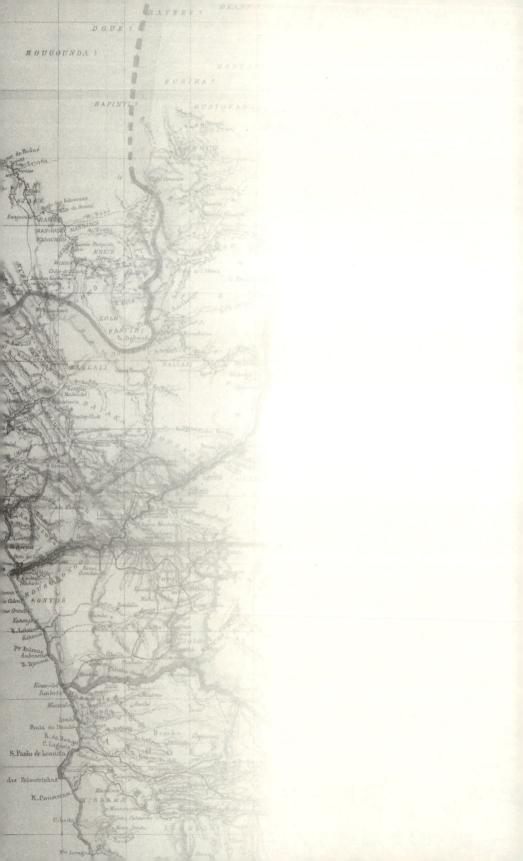

HELP SUPPORT AFRICAN HERPETOLOGY RESEARCH

As noted in this book, funding sources that support African biodiversity surveys are increasingly scarce, and the remaining ones are extremely competitive. If you would like to make a gift to support the type of research described in this book, please use the QR code or website below to make a donation. Funds will support historically underrepresented students at the University of Texas at El Paso, including Hispanics and women, and students, biologists, and protected area rangers in African countries. Thank you for your support!

https://givingto.utep.edu/Herpetology

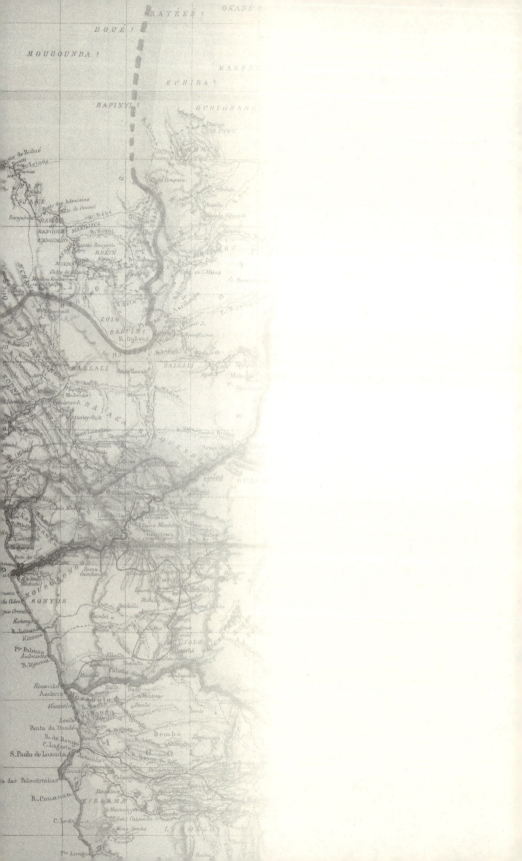

Preface

1. Here and elsewhere in the book, I use Congo as a shorthand for Democratic Republic of the Congo. When the neighboring Republic of Congo is mentioned, it is not abbreviated.

2. The longest river in the world is the Nile.

3. Giam 2017.

4. Mora et al. 2011.

5. Rosa et al. 2016; J. Graham 2022.

6. Greenbaum 2017. See chapter 1 of this book for a primer on Congo's history, including surprising links to the United States.

7. Sage 2020; J. Graham 2022.

8. Brander et al. (2024:2) defines ecosystem services as "the direct and indirect contributions of nature to human well-being." These services include, just to name a few, water filtration, precipitation and temperature regulation for local and regional areas, flood prevention, erosion reduction, pest and disease control, and a sustainable source of food, pollinators, medicine, and construction materials (Sodhi et al. 2007; Brander et al. 2024).

9. Oliveira et al. 2022; Tingley 2024.

10. Ghazoul and Sheil 2010.

11. Howes et al. 2020.

12. S. H. Lee et al. 2021.

13. Re:wild et al. 2023.

14. G. Einhorn and de Merode 2025.

15. G. Li et al. 2024.

16. Greenbaum 2017.

Chapter One

1. See Henderson and Nicholls (2015). The paper's jaw-dropping "Double Death" (fig. 1) depicts two *C. saharicus* carrying away a bleeding *Limaysaurus tessonei* victim after a successful sunset ambush of a herd of the sauropods.

2. O'Hanlon 1997.

3. Pronounced *wan-day-gay*.

4. One of the largest families of flowering plants in the world, the Euphorbiaceae (spurge family) includes the colorful and popular croton (*Codiaeum variegatum*) houseplant, which is native to the Australasian tropics (Brenzel 2007; PoWO 2022).

5. Chifundera informed me that he is studying the phytomolecules of this plant in his research laboratory in Kinshasa. He says pharmacological studies are ongoing with World Health Organization scientists (C. Kusamba, in litt., August 13, 2022).

6. Portillo et al. 2019.

7. In this book I follow the definition of *venom* as defined by Jenner and Undheim (2017:12): "Venom is broadly defined as a toxic secretion produced by specialized cells in one animal that is delivered to another animal via a delivery mechanism—typically through the infliction of a wound—to disrupt normal physiological functioning in the interest of predation, feeding, defense, competition or other biological processes that benefit the venom-producing animal." By this definition, mosquitoes are venomous, but animals such as the famous Neotropical poison dart frogs that passively transfer toxins via skin secretions are poisonous.

8. Deufel and Cundall 2003.

9. Warrell 1995; P. Wagner et al. 2009; Oulion et al. 2018.

10. Loveridge 1938.

11. Ehret 2002; Fromont 2023.

12. Fromont (2023) noted this name means "King of the kingdom."

13. Balandier 1968.

14. Lagamma 2016.

15. Balandier 1968.

16. Fromont 2023.

17. Balandier 1968.

18. Lagamma 2016.

19. Balandier 1968. Fromont (2023) noted that the head was taken to Luanda, where it was supposedly displayed on the wall of the Church of Our Lady of Nazareth.

20. Lagamma 2016.

21. Britannica 2007; Sautter and Pourtier 2025.
22. Pocock drowned soon afterward, and thus we must take Stanley's word for it (Stanley 1878/1988).
23. As noted by Sautter and Pourtier (2025), it is "virtually certain" that seventeenth-century Capuchin missionaries reached Malebo Pool.
24. Stanley 1878/1988:255. By this time, the Teke people, associated with the Tyo kingdom, seem to have displaced their ancient rivals of the Kongo kingdom from the southern side of Malebo Pool (Ehret 2002). However, the Kongo language is still spoken today at Malebo Pool (Lagamma 2016).
25. Stanley 1878/1988:329.
26. Penn 2001; Britannica 2007; Van Reybrouck 2014.
27. According to Van Reybrouck 2014:49, Kinshasa was known as "Paradise of the Pool" in 1884.
28. Stuart 1975.
29. At the time of Congo's independence in 1960, the country had more hospital beds than all other "black African countries" combined, according to Wrong (2002).
30. Anonymous 1956.
31. Mitchell 2014.
32. Silva 2017.
33. McKenna 2011.
34. Ledgard 2014; Cordell et al. 2025.
35. Vidal 2018; Bédécarrats et al. 2019.
36. Ledgard 2014; Anonymous 2025a.
37. Roth and Sawyer 2017; Freytas-Tamura 2018, 2019.
38. Guy-Crispin Gembu Tungaluna (in litt. March 24, 2022) via correspondence with Erik Verheyen. See also Van Cakenberghe et al. (2017).
39. Rookmaaker 2013.
40. Roosevelt 1910/1987; Edwards 2016; Hsu 2017.
41. Hsu 2017; Neme 2017; Nam Dang Vu and Reinhardt Nielsen 2018.
42. Emslie 2020a. This organization was founded by British evolutionary biologist Julian Huxley as the International Union for the Protection of Nature in 1948, and it transitioned to the International Union for the Conservation of Nature and Natural Resources (IUCN) in 1956 ("About IUCN," IUCN, accessed February 3, 2025, https://iucn.org/about-iucn; Mann 2018).
43. Goyanes 2017.
44. Emslie 2020a.
45. Neme 2017; Winsor 2018.

46. For a discussion of the taxonomic and conservation ramifications of subspecies vs. species, see Greenbaum (2017).

47. Kumar et al. 1993.

48. Harley et al. 2016.

49. Rookmaaker 2013.

50. Saragusty et al. 2016; Sánchez-Barreiro et al. 2021.

51. Nuwer 2018. Hillman Smith (2014:270) noted that the rhino search "could account for four different individuals" in 2006, even though only one animal was observed.

52. Harley et al. 2016.

53. Ibid.; Nuwer 2018.

54. Fakhri 2018.

55. Saragusty et al. 2016; Hildebrandt et al. 2018; Busby 2022. In 2023–2024 a consortium of European and African scientists and wildlife biologists called BioRescue made significant advances with this effort by producing viable southern white rhino embryos from in vitro fertilization. The hope is that the new methods can be used for the northern white rhino (Gretener 2024).

56. Kluger 2024.

57. Charlesworth and Willis 2009.

58. Yang et al. 2024.

59. Sánchez-Barreiro et al. 2021.

60. Although many technical challenges would need to be addressed before this could be possible, scientists working for a company called Colossal Biosciences are hard at work trying to use genomic data from the extinct woolly mammoth (*Mammuthus primigenius*) to bring it back from extinction (Neuman 2021; Busby 2022), and similar methods could be used to restore more recently extinct species. In 2024, Colossal announced it had created induced pluripotent stem cells for Asian elephants, the closest living relative of the woolly mammoth. The advancement is another step forward to resurrect the mammoth (Stein 2024). Later that year, scientists published a study describing the 3-D structure of chromosomes from a 52,000-year-old sample of mammoth skin found in Siberian permafrost (Guglielmi 2024).

61. Emslie 2020b.

62. Anonymous 2023.

63. Wrong 2002.

64. Leaché et al. 2017; Krishnan et al. 2019.

65. Nuñez et al. 2016.

66. In fact, TSD and genetic-sex determination occur on a continuum whereby both environmental and genetic factors can contribute to the sex of offspring of a given species, but this can vary by population and even family groups within populations (Steele et al. 2018).

67. Ghazoul and Sheil 2010.

68. Kayumba et al. 2015; Fondation Bombo Lumene website, accessed February 3, 2025, https://fblu.org/.

69. Bauer et al. 2015, 2022.

70. Agarwal et al. 2021; Uetz et al. 2025.

71. Weterings and Vetter 2018.

72. Allen et al. 2019; Agarwal et al. 2021.

73. Tolley et al. 2016; Greenbaum 2017; Farooq et al. 2021.

74. Fondation Bombo Lumene website, accessible at https://fblu.org/.

75. Balandier 1968.

76. Milau et al. 2016.

77. Evans et al. 2015, 2019. Most *Xenopus* are known for their multiple sets of chromosomes (i.e., polyploidy), which originated during multiple independent events during their evolutionary history and diversification. This complexity is especially salient in the evolution of their sex chromosomes (Cauret et al. 2020; Evans et al. 2024).

78. Ceríaco et al. 2021. In a study focused on Angolan geckos of this genus (Lobón-Rovira et al. 2021), some of my colleagues cast doubt on the validity of this species, a fair critique given the small sample size (N = 2) in the original study that described *Hemidactylus gramineus*. The sequence divergence (8.42 percent in the ND2 gene) they found between *H. gramineus* and its closest relative in Angola (*H. nzingae*) is lower than several other similar comparisons between closely related species, and yet it is far higher than measures of intraspecific (among populations of the same species) variation in many species of geckos in this genus. Two subsequent studies from the same journal described new species of *Hemidactylus* that had ND2 sequence divergences that were even lower—closer to 6 percent (e.g., Das et al. 2022; Narayanan et al. 2023). Moreover, a study that produced a phylogeny of South Asian *Hemidactylus* geckos suggested a species-level cutoff of 5 percent sequence divergence for the ND2 gene (Agarwal et al. 2019). Scientists by nature are skeptical, this interesting taxonomic case certainly has gray areas, and I welcome additional scrutiny to clarify the taxonomic status of *H. gramineus*.

79. Sage 2020.

80. Anonymous 2020.

Chapter Two

1. Maurel 1992.

2. Galloway 2021.

3. Böhme 2000.

4. Known from a single male specimen collected in the remote village of Ikela in 1957, *Mehelya laurenti* was named by Belgian herpetologist Gaston-François de Witte to honor his fellow countryman and herpetologist Raymond Laurent (de Witte 1959). This species is likely rare, because it was never encountered again (Broadley et al. 2018; Chippaux and Jackson 2019).

5. Chaney et al. 2024.

6. Figueroa et al. 2016; Burbrink et al. 2020.

7. Pauwels et al. 2002; Chippaux and Jackson 2019.

8. Pauwels et al. 2002.

9. According to Richie et al. (2020), about 48 percent of the people in Republic of the Congo have access to electricity, whereas in Democratic Republic of the Congo the estimate is only 19 percent. This disparity was similar in 2013 during the expedition.

10. Kirstein et al. 2013; Bébé Ngouateu and Dondji 2022.

11. Bébé Ngouateu and Dondji 2022.

12. Aronson et al. 2017.

13. Alvar et al. 2021; Kumar Tripathi and Kumar Nailwal 2021; Bébé Ngouateu and Dondji 2022.

14. De Waal and Lanting 1997.

15. Quammen 2013.

16. Inogwabini et al. 2008. These high bonobo population estimates were recon-firmed by Inogwabini (2020).

17. Inogwabini et al. 2008.

18. Inogwabini (2020) noted the presence of lions at Malebo, about ten miles north of Nkala, as recently as 2008. One person was killed by these lions, suggesting the bonobos would be wise to keep an eye out for cat predators, including leopard as well.

Chapter Three

1. Nürk et al. 2020.

2. De Waal and Lanting 1997; Diogo et al. 2017; Wrangham 2023.

3. Owen-Smith 2021.

4. Werdelin and Peigné 2010; Siliceo et al. 2020.

5. Zachos et al. 2001; Kissling et al. 2012.

6. Sepulchre et al. 2006; Linder 2017.

7. Leakey et al. 1996; Kissling et al. 2012.

8. Langergraber et al. 2012; Diogo et al. 2017; Besenbacher et al. 2019; Zimmer 2022.

9. Owen-Smith 2021. *Australopithecus afarensis*, a possible human ancestor made famous by the fossil dubbed Lucy, may have taken its first bipedal steps in a grassy woodland environment about 3.85 million years ago (Gibbons 2024). However, Roberts (2022) suggested bipedalism might have evolved in tropical forests.

10. Prüfer et al. 2012; De Manuel et al. 2016; Mao et al. 2021; Wrangham 2023.

11. Takemoto et al. 2015.

12. De Manuel et al. 2016.

13. I must credit this tongue-in-cheek comment to Carl Lieb, emeritus professor of herpetology at University of Texas at El Paso.

14. De Waal and Lanting 1997.

15. De Waal and Lanting 1997; Woods 2010; Wrangham 2023.

16. Baisas 2023.

17. Wrangham 1993; de Waal and Lanting 1997; Greenwood 2007; Quammen 2013.

18. See Wrangham (2023) for another hypothesis involving mutation-order speciation.

19. Southern et al. 2021.

20. Wrangham 1993; Groves 2013; Quammen 2013; Tocheri et al. 2016. Wrangham (2023) explores the idea of "self-domestication" in bonobos to explain their docile nature.

21. Ordaz-Németh et al. 2021; Stirn 2021.

22. Humle et al. 2016; S. Heinicke et al. 2019.

23. Fruth et al. 2016.

24. Barthel et al. 2022. Another source (Megevand 2013) approximates the Congo Basin forests at about 300 million hectares. The basin extends across parts of Cameroon, Central African Republic, Democratic Republic of the Congo, Republic of the Congo, Equatorial Guinea (mainland), and Gabon.

25. A German duke who was traveling down the Congo River toward Léopoldville in 1908 remarked in passing that the telegraph line connecting the city to

Coquilhatville was frequently severed by the destructive nature of elephants (Frederick 1910).

26. Hochschild 1999.

27. Van Schuylenbergh 2019.

28. Conrad 1899/1980:33.

29. Anonymous 1956.

30. Greenwood 2007. Luckily, the war never reached Nkala, which might explain the site's large bonobo population (Inogwabini 2020).

31. Pough et al. 2016; Spawls et al. 2018; Chippaux and Jackson 2019.

32. Trape and Mané 2006.

33. Greene 1997; Spawls et al. 2018.

34. Christy 1924:26.

35. This call can be heard at the following website: "*Corythaeola cristata*," eBird, accessed February 3, 2025, https://ebird.org/species/grbtur1.

36. Dyck 1992; Vernon and Winney 2000.

37. Chapin 1939; Candy 1984; Turner and Kirwan 2020.

38. BirdLife International 2017; Turner and Kirwan 2020.

39. This group of insects includes katydids.

40. Bateman and Fleming 2009.

41. Nok 2009.

42. Lyons 2002; Malvy and Chappuis 2011.

43. As discussed by Alsan (2015), advanced civilizations like Great Zimbabwe thrived in areas where the tsetse fly was absent.

44. Diamond 1999.

45. Malvy and Chappuis 2011; Dickie et al. 2020; Falisse and Mpanya 2022.

46. Fairlamb and Horn 2018.

47. Tkach and Columbia Broadcasting System 2001; Malvy and Chappuis 2011.

48. Dickie et al. 2020; Falisse and Mpanya 2022.

49. Falisse and Mpanya 2022.

50. Ibid.

51. Hemingway 1935/2015:94.

52. Ferrari et al. 2016; Deutsch-Feldman et al. 2021.

53. Emina et al. 2021.

54. I must credit my colleague David Blackburn, professor at the Florida Museum of Natural History, for teaching this technique to us during an expedition to Burundi in 2011.

55. Hirschfeld et al. 2015.

56. Blackburn 2008.

57. Blackburn 2009.

58. This quote was translated by Blackburn (2009), who is the world authority on arthroleptid frogs.

59. Bittencourt-Silva et al. 2020.

60. In 2011, my Congolese colleagues and I joined herpetologist David Blackburn on a quest to find the Bururi long-fingered frog (*Cardioglossa cyaneospila*) in Burundi. Despite our combined decades of fieldwork experience and hours of searching, only David managed to find a single individual near a creek where dozens of males were calling from the grassy vegetation at our feet (Blackburn et al. 2016).

61. Blackburn et al. 2021; Badjedjea et al. 2022.

62. Jenner and Undheim 2017.

63. Shubin 2008; Pough et al. 2016; Mullen and Sissom 2019.

64. Mullen and Sissom 2019.

65. I am grateful to Lorenzo Prendini, curator of arachnids at the American Museum of Natural History in New York, who identified both scorpions mentioned in this book.

66. Prendini 2015.

67. Newlands and Martindale 1980.

68. Mullen and Sissom 2019. One component of this scorpion's venom, chlorotoxin, is used as a marker for tumors in the central nervous system (Jenner and Undheim 2017).

Chapter Four

1. Bates et al. 2013; Pough et al. 2016; Spawls et al. 2018.

2. C. Williams et al. 2022.

3. Engelbrecht et al. 2019.

4. Spawls et al. 2018.

5. Dowell et al. 2016; Spawls et al. 2018.

6. Spawls et al. 2018.

7. Compere and Symoens 1987.

8. I want to acknowledge that some people, mostly in academia, object to the term *American* for citizens of the United States, because they say it implies that the United States is the most important country in the New World (J. Lee 2022). Let me be clear that I do not necessarily think the United States is more important than other countries in the Western Hemisphere, but I decline to adopt Stanford's replacement term of "US citizen" here.

9. Stewart and Roberts 1984.

10. Ibid.; Moelants 2010.

11. *Mai-Ndombe* means "dark water." The color of the water is likely linked to nutrient-poor and acidic properties of the lake (Ghazoul and Sheil 2010).

12. Grenfell explored over 12,000 miles of the Congo and several other rivers (Van Reybrouck 2014).

13. Johnston 1908.

14. Vansina 1990.

15. Stengers 1957; De Roo 2017; Van Schuylenbergh 2019.

16. Van Reybrouck 2014.

17. Hochschild 1999.

18. Galloway 2021.

19. Rorison (2008) noted that Congo is where American T-shirts go to die. Because the vast majority of Congolese people do not speak English, they usually do not know what the shirts say and cultural references are completely lost. They are selected for their color, creative artwork, or cheap price. Many of these unwanted shirts were discarded by their American owners because they could not be worn in public. More than once, I have seen Congolese people donning shirts that feature curse words or crude bodily functions.

20. Chapin 1939; Ghazoul and Sheil 2010.

21. Kirwan et al. 2021.

22. Ghazoul and Sheil 2010.

23. Ghazoul and Sheil 2010; Parolin and Wittmann 2010.

24. Based on 4 males collected in northern Angola, this species was described as *Hemidactylus pfindaensis* in 2021. We later noticed that DNA from the specimen found at Mpote-Emange matched the Angolan specimens, and the range of the species was thus extended by over 375 miles from the nearest record (Greenbaum et al., 2025).

25. Daley 2016; Prat et al. 2016.

26. Anthes 2022.

27. Happold (2013:223) noted a lack of consensus about the number of species in Africa.

28. Cornell Lab Macaulay Library website, accessed February 3, 2025, www.macaulaylibrary.org/.

29. Several recordings from Gabon in the early 1970s by Jack Bradbury.

30. Happold and Happold 2013.

31. The males calling at Mpote-Emange seemed to be perched in a fig tree, but it is unclear if the tree had fruit at the time of my visit in late June 2013.

32. In West African populations that are putatively the same species, a harem-style system is in place (Olson et al. 2019).

33. Ibid.

34. Bawa 1990.

35. Quammen 2012; Gorman 2024.

36. K. Jones et al. 2008; Karesh et al. 2012.

37. Van Seventer and Hochberg 2017.

38. Quammen 2012, 2014; J. Hawkins et al. 2019; Olson et al. 2019.

39. Leroy et al. 2009; Ripple et al. 2019.

40. Cohen 2022. It is possible SARS-CoV-2 first infected humans via an intermediate host at the Huanan Seafood Wholesale Market in Wuhan City, China. The coronavirus has been shown to infect hamsters, mink, and white-tailed deer, suggesting "host generalist capacity" (Pekar et al. 2022). However, recent investigations suggested it is also possible the virus originated from the Wuhan Institute of Virology, where coronavirus research was occurring in 2019 (Eban and Kao 2022; Barnes 2025).

41. Afelt et al. 2018.

42. Wallace-Wells 2022.

43. Rosa et al. 2016:fig. 1.

44. Paice 2022.

45. Kumakamba et al. 2021; Shapiro et al. 2021.

46. Testing in Kinshasa suggested the pathogen is not Ebola or Marburg virus, but some of the hemorrhagic fever symptoms are similar. Victims often died within 48 hours and the fatality rate was estimated to be over 12 percent (Ho 2025).

47. Craig 2024. About one in ten people who become infected with this new strain of Mpox will die, and the vast majority of new cases are children under the age of fifteen (Emanuel 2024).

48. Craig 2024; Ringstrom and Steenhuysen 2024.

49. Cuevas 2024.

50. LeBreton et al. 2007.

51. I must thank Roy Gereau (Missouri Botanical Garden) and Andy Plumptre (BirdLife International, Cambridge, UK) for confirming that this tree is a fig (*Ficus*). There is not enough detail in the photo I took to narrow it down to species.

52. Ghazoul and Sheil 2010; Pothasin et al. 2014; Clement et al. 2020.

53. Ghazoul and Sheil 2010; Clement et al. 2020.

54. However, as shown by Wang et al. (2021), frequent host switching by the wasps has likely led to interspecific hybridization in the figs during their coevolutionary history.

55. Bawa 1990; Shanahan et al. 2001; Ghazoul and Sheil 2010; Pothasin et al. 2014; Clement et al. 2020.

56. Jangannathan and Kakuru 2022. Several malaria vaccines are on the horizon, and some experts believe it will be possible to eradicate the disease in a few years. However, many challenges remain. One survey found that half of the people in Congo would not trust a malaria vaccine (Mandavilli 2022). Another huge challenge is funding. Malaria vaccines cost millions of dollars to develop and test, but the people who need them the most cannot afford to pay for them (Nolen 2024).

57. Kingdon 2023:200.

58. Chauhan et al. 2017; K. Tan and Arguin 2017; Bogitsch et al. 2019.

59. Orish et al. 2019.

60. Kariuki and Williams 2020.

61. Russell 2010; Pierce 2020.

62. Ibid.

63. Solovieff et al. 2011.

64. Schäfermann et al. 2020.

65. I am grateful to my colleague Václav Gvoždík for this identification (in litt., November 10, 2024).

66. Rabb and Rabb 1963; Pearl et al. 2000.

67. Hime et al. 2021; Frost 2025.

68. Spawls et al. 2018.

69. Weinstein et al. 1991.

70. Confirmed by Scott Weinstein (in litt., August 22, 2022).

71. Swaroop and Grab 1954.

72. Loveridge 1931.

73. Dobiey and Vogel 2007.

74. Chippaux and Jackson 2019.

75. Wolfgang Wüster (in litt., April 7, 2020); Tomáš Mazuch (in litt., July 22, 2022).

76. Zoltán Nagy (in litt., May 28, 2014).

77. Collet and Trape 2020.

78. Coates et al. 2018; Václav Gvoždík (in litt., May 19, 2022).

79. See Greenbaum (2017) for a discussion of subspecies and taxonomic classification.

80. Montilly 2021.

81. Muiruri 2022.

Chapter Five

1. Search from August 27, 2022.

2. Homer 1712; Verdu 2016. As noted by Turnbull (1968:15) and others, Homer may not be the earliest reference. A document from the fourth dynasty of Egypt notes how a commander named Herkouf (a.k.a. Harkhuf) entered a forest west of the Mountains of the Moon (Rwenzori Mountains on the Congo/Uganda border) and encountered "a people of the trees, a tiny people who sing and dance to their god."

3. Verdu 2016; Hewlett 2017.

4. Liebowitz and Pearson 2005; Jeal 2007; Murray 2012.

5. Turnbull 1968; Anonymous 2015; Verdu 2016; Greenbaum 2017; Hewlett 2017; Strochlic 2017.

6. Verdu 2017.

7. At five feet, ten inches tall, I can attest that my head is constantly getting ensnared in lianas, branches, and creepers when I try to move through the dense vegetation of the rainforest.

8. Verdu 2017.

9. Klieman 2003.

10. Ballard 2006.

11. Bradford and Blume 1992; Keller 2006.

12. This term is also a generalization that encompasses a diverse array of African groups. Hewlett (2017) points out that *Bantu* is a linguistic term and there are many groups in the Congo Basin who speak non-Bantu languages. However, *Bantu* seems to be a commonly used name to describe non-pygmy groups (e.g., Vansina 1990; Diamond 1999; Klieman 2003).

13. Hewlett 2017.

14. Vansina 1990; Diamond 1999; Patin et al. 2014; Lupo et al. 2017; Lorente-Galdos et al. 2019.

15. Klieman 2003.

16. Ballard 2006.

17. Fa et al. 2016.

18. Rupp 2011.

19. Schweinfurth 1874; Klieman 2003; Di Campo 2017; Hewlett 2017.

20. Hewlett 2017.

21. Maclean and Kabanda 2022.

22. Verdu (2016) notes there are no traces of an ancient pygmy language, but Diamond (1999:368) suggests traces of their original language are evident in "some words and sounds."

23. According to Verdu (2016), there might be as many as 350,000. A study using spatial distribution models by Olivero et al. (2016) suggested there is enough favorable habitat to support 900,000 hunter-gatherer people in Central African forests.

24. Hewlett 2017.

25. Even this name might have problematic origins. According to an American missionary doctor, Jerry Galloway, the Batwa received their name from the Ekonda. Galloway referred to the former group by their original name, the Botoa (Galloway 2021).

26. Chabiron et al. 2013.

27. Turnbull 1968.

28. Vansina 1990; Klieman 2003; Verdu 2017.

29. Verdu 2017.

30. Patin et al. 2014; Verdu 2017.

31. Verdu 2017; Choudhury et al. 2020.

32. Hallet 1967a; Galloway 2021.

33. The discrepancy in spelling of this village might be due to a relatively common, colonial-era error of transliteration by the Belgians.

34. See chapter 3 of Greenbaum (2017) to understand how we learned this lesson the hard way.

35. Identified by Piet Stoffelen, senior researcher at Meise Botanic Garden, Belgium on February 28, 2019.

36. Groom 2012; Ahossou et al. 2020.

37. Hopkins 1983; Hopkins and White 1984.

38. Tchinda 2008.

39. Schäfer et al. 2022; Frost 2025.

40. Nagy et al. 2014.

41. Kusamba et al. 2013.

42. Portillo et al. 2018, 2019.

43. Weinstein et al. 2021.

44. De Witte and Laurent 1943.

45. I follow the Reptile Database (Uetz et al. 2025) and the Crocodile Specialist Group (Matthew Shirley, in litt., April 18, 2023) in my recognition of *Osteolaemus osborni*.

46. Spawls et al. 2018.

47. Crocodile Specialist Group 1996.

48. Joris 1992:137.

49. Zoer 2012; Gore et al. 2021.

50. Anonymous 1956.

51. Bampton 1962.

52. Schiøtz 2006; Rödel et al. 2019.

53. Knight 2003; Olivero et al. 2016.

54. Sodhi et al. 2007; Olivero et al. 2016.

55. A recent study in the journal *Nature* found that negative edge effects (e.g., exposure to desiccating wind, microclimate differences, proximity to human activities) can penetrate over a mile into the forest in Africa and even farther in the Neotropics (Bourgoin et al. 2024).

56. Cuthill et al. 2000; Brecko and Pauwels 2024.

57. Schiøtz and Volsøe 1959; Arnold 2002.

58. Vanhooydonck et al. 2009.

59. Arnold 2002; Spawls et al. 2018.

60. Kingdon and Hoffmann 2013a.

61. Gossé et al. 2024.

62. Kingdon and Hoffmann 2013a.

63. Kehinde Omifolaji et al. 2022.

64. Zhang et al. 2020.

65. Kehinde Omifolaji et al. 2022.

66. Akani et al. 2013.

67. Chippaux and Jackson 2019.

68. Meier and White 1995.

69. Hernandez 2022.

70. G. Li et al. 2024.

71. Isbell 2006; R. Harris et al. 2021.

72. McCleary and Kini 2013; Oliveira et al. 2022.

73. Bhatt et al. 2022.

74. Oliveira et al. 2022.

75. Wilcox 2016; Jenner and Undheim 2017; Prater 2023; Tingley 2024.

76. Tingley 2024.

77. Oliveira et al. 2022.

78. Greenbaum et al. 2003; Oliveira et al. 2022.

79. Casewell et al. 2020. Some estimates for snakebite victims are as high as 5.5 million per year (Tingley 2024).

80. Originally described as *Toxicodryas adamanteus*, but later corrected to *T. adamantea* because the genus name is feminine (Greenbaum et al. 2021; Pauwels and Colyn 2023).

81. Grossman 2019.

82. Dargie et al. 2017; Craft 2022.

83. IUCN 2021; Maclean and Kabanda 2022.

84. Dargie et al. 2017; Maclean and Kabanda 2022.

85. Crezee et al. 2022. At least some of these peat areas seem to occur in the proposed Kivu-Kinshasa Green Corridor I mentioned in the preface (G. Einhorn and de Merode 2025).

86. Zarin 2022.

Chapter Six

1. Emerson 1979.

2. McLynn 2004.

3. Liebowitz and Pearson 2005.

4. Anonymous 1956:646.

5. Devlin 2007.

6. Hunt 2016:252.

7. Mangulu 2003.

8. Wrong 2002.

9. Hunter 2016:21.

10. Ibid.

11. Mangulu 2003.

12. Stearns 2011.

13. Anonymous 1956.

14. "Mbandaka, Republic of Congo Metro Area Population 1950–2025," macrotrends, February 3, 2025, www.macrotrends.net/cities/20858/mbandaka/population.

15. Vande weghe and Vande weghe 2018.

16. Curry-Lindahl 1974; J. Hart and Hall 1996; Van Krunkelsven et al. 2000.

17. Fonteyn et al. 2023.

18. Van Krunkelsven et al. 2000; Fonteyn et al. 2023. This World Heritage Convention was recognized in 1972 to "protect natural and cultural heritage of outstanding universal value." *Natural heritage* refers to areas with exceptional aesthetic, physical, biological, or geological characteristics, including habitats that are essential for conservation of threatened plant and animal species (Debonnet and Hillman-Smith 2004).

19. Browne 2021.

20. Van Krunkelsven et al. 2000; Furuichi and Thompson 2008; Mulotwa et al. 2010; Bessone et al. 2020.

21. Vande weghe and Vande weghe 2018; Jo Thompson and Thomas Butynski (in litt. July 25, 2023).

22. Laurent 1965; Broadley 1998.

23. Schiøtz 1982:261.

24. Kielgast and Lötters 2009.

25. Schick et al. 2010.

26. Anonymous 1964.

27. Maurel 1992.

28. Lobón-Rovira et al. 2024. This study was originally published online in 2023.

29. Lorenzo Prendini (in litt., September 23, 2013).

30. Lauren Esposito (in litt., July 31, 2023).

31. Stockmann and Ythier 2010.

32. H. Tan and Mong 2013.

33. Hallet 1967b.

34. Arnold 1989.

35. Schmidt 1919.

36. Frost 2025.

Chapter Seven

1. Uetz et al. 2025.

2. Benton and Clark 1988; Butler et al. 2011; Stevenson 2022.

3. Pough et al. 2016.

4. Yoshida et al. 2021; Stevenson 2022.

5. Pough et al. 2016; Stevenson 2022.

6. Stevenson 2022.

7. Ibid.

8. Faaborg 2020.

9. Rico-Guevara et al. 2022; Schlinger 2023.

10. Stevenson 2022; Ham 2023; Kilvert 2023.

11. Shirley et al. 2018.

12. Faaborg 2020.

13. This also happens in the South American river turtle (*Podocnemis expansa*) (Pough et al. 2016).

14. Pough et al. 2016; Stevenson 2022.

15. Stevenson 2022.

16. Anonymous 2014.

17. Foster and Foster 2013.

18. Chavan and Borkar 2023.

19. Stevenson 2022.

20. Izadi 2016.

21. IUCN Red List website, accessed April 12, 2024, www.iucnredlist.org; Stevenson 2022.

22. Shirley et al. 2014.

23. A record from Uganda is doubtful (M. Shirley, in litt., November 7, 2023).

24. Eaton et al. 2009; Ceríaco et al. 2018; Britz et al. 2020.

25. A. Graham and Beard 1990; Spawls et al. 2018; Stevenson 2022.

26. Hekkala et al. 2011.

27. Shirley et al. 2018.

28. Ibid.

29. Spawls et al. 2018.

30. Frost 2025.

31. Greenbaum et al. 2012; Greenbaum et al. 2025.

32. Spawls et al. 2018.

33. Spawls and Branch 2020.

34. Allen et al. 2019.

35. Chippaux and Jackson 2019.

36. Spawls and Branch 2020.

37. This locality was transliterated as Lotulu in the author's field notes from 2013 and reported by this name in Nečas et al. (2024). Based on maps in Vande weghe and Vande weghe (2018), the spelling was modified herein to Lotulo.

38. Zhou et al. 2018.

39. Identification confirmed by V. Gvoždík on May 28, 2023; Nečas et al. 2024; Frost 2025.

40. Blackburn et al. 2022; Greenbaum et al. 2022.

41. Stevenson 2022.

42. Browne 2021.

43. Ibid.

44. Correa et al. 2024.

Chapter Eight

1. Crawford 2018; Wilson 2022.

2. There are a growing number of recently discovered, notable exceptions, including the creatively named pandoraviruses, medusaviruses, mimiviruses, and pithoviruses (Schulz et al. 2022).

3. Crawford 2018; Wilson 2022.

4. Recent publications (e.g., Gribaldo and Brochier-Armanet 2019) suggested there should be only two domains of life, because eukaryotic organisms evolved within a lineage of Archaea. Previous phylogenetic analyses may have been misguided by long branch lengths.

5. Wilson 2022.

6. Marr 2007.

7. There are probably additional reasons, including the double-helix structure of DNA (Pierce 2020).

8. Crawford 2018; Page 2023.

9. Timberg and Halperin 2012; Pépin 2021.

10. Quammen 2014; Masci and Bass 2018; Pépin 2021.

11. Masci and Bass 2018.

12. Alimohamadi et al. 2021.

13. Masci and Bass 2018.

14. Qureshi 2016; Farmer 2021; Daniel 2024.

15. Masci and Bass 2018.

16. Ibid.

17. Daniel 2024.

18. Redding et al. 2019.

19. Spawls et al. 2018.

20. Spawls and Branch 2020.

21. According to Richard's brother Robert Blum (in litt., July 15, 2011), his great-grandfather on his mother's side was Louis Nathan Heil, who was the brother of my great-great-grandfather, Abraham Heil.

22. Diana Natalicio, in litt., August 2, 2013. Diana Natalicio was beloved at UTEP. She passed away in September 2021.

23. Heymann 2014; Nanclares et al. 2016.

24. Starin and Burghardt 1992; Fredriksson 2005; Spawls et al. 2018; Murphy and Crutchfield 2019.

25. Du Chaillu 1861; Loveridge 1931.

26. Bentley 1900:393.

27. Kingsley 1897:546; Murphy and Crutchfield 2019.
28. Curran and Kauffeld 1937; Luiselli et al. 2007; Spawls et al. 2018.
29. Spawls et al. 2018.
30. Greene 1997.
31. Schweinfurth et al. 1888; Davis 1940; Joy 1951; Greene 1997.
32. Richardson 1972.
33. Schweinfurth et al. 1888.
34. Spawls et al. 2018.
35. Lyons 2002.
36. Brian Weiss, in litt., January 27, 2020.
37. Portillo et al. 2018. In 2025, Frank Portillo, my former PhD student, was preparing a publication to describe this snake as a new species.
38. De Witte 1965.
39. Beolens et al. 2011; Tilbury 2018.
40. Tilbury 2018.
41. Tolley and Herrel 2013.
42. Mocquard 1906.
43. De Witte 1965; Tilbury 2018.
44. Tolley and Herrel 2013.
45. Karsten et al. 2008.
46. Nicholson et al. 2018.
47. Tolley and Herrel 2013; Anderson et al. 2015; Díaz and Malhi 2022.
48. Kingdon and Hoffmann 2013b.
49. Ralls 1978.
50. Kingdon and Hoffmann 2013b.
51. Kingdon and Hoffmann 2013b; IUCN SSC Antelope Specialist Group 2016.
52. Ripple et al. 2019.
53. Benítez-López et al. 2019; Leisher et al. 2022.

Chapter Nine

1. Marsh et al. 2025.
2. Mukpo 2022; Oakland Institute 2024.
3. Ibid.
4. Anonymous 2025b.
5. McDowell and Mason 2020.
6. T. Li 2018; Meijaard et al. 2020.

7. Vijay et al. 2016; Meijaard et al. 2020.

8. According to their website, "Trase is a not-for-profit initiative founded in 2015 by the Stockholm Environment Institute and Global Canopy to bring transparency to deforestation and agricultural commodity trade." Trase website, accessed February 18, 2025, https://trase.earth/.

9. Anonymous 2024a.

10. Mukpo 2022.

11. Oakland Institute 2024.

12. Like many other widespread species of amphibians and reptiles in Central Africa, genetic analyses suggest *Hylarana albolabris* is a complex of cryptic species, and the populations from western Congo are likely a new species (Jongsma et al. 2018).

13. Stearns 2011.

14. Van Reybrouck 2014. About 80 percent of the refugees were women and children (Umutesi 2004).

15. Ibid.

16. In 2024, five new cryptic species that had been masquerading under the name *Rhampholeon boulengeri* were described (Hughes et al. 2024).

17. Tilbury 2018; Burbrink et al. 2020.

18. Beolens et al. 2011. One of the world authorities on trionychid turtles was my UTEP colleague Robert Webb, who passed away in 2018 at age 91. See Greenbaum and Zhuang (2019) for a summary of his incredible career in herpetology.

19. Pough et al. 2016.

20. Turtle Taxonomy Working Group 2021; J. L. Carr, in litt., July 17, 2024.

21. Gramentz 2008; Diagne et al. 2013.

22. Gramentz 2008.

23. Gramentz 2008.

24. Chirio et al. 2017.

25. Gramentz 2008.

26. Chirio et al. 2017.

27. Diagne et al. 2013.

28. Vargas and Zardoya 2014; Stanford et al. 2020.

29. Stanford et al. 2020.

30. Stanford et al. 2020; Sullivan 2023.

31. Bonin et al. 2006; Bryan 2023.

32. Stanford et al. 2020.

33. Ibid.

Chapter Ten

1. Branch 1984; Murphy and Henderson 1997.

2. Minton and Minton 1973.

3. Greene and Wiseman 2023.

4. Murphy and Henderson 1997.

5. Buel 1890:262. In the nineteenth century and beyond, white explorers and colonists often referred to adult African men as "boys" (e.g., Morrell 1998). In his posthumously published book *Under Kilimanjaro*, Hemingway (2005:8) seems to reflect on this when he remarks, "Once they had been the boys. They still were to Pop. But he had known their fathers when their fathers were children. Twenty years ago I had called them boys too and neither they nor I had any thought that I had no right to. Now no one would have minded if I had used the word. But the way things were now you did not do it."

6. Stanley 1878/1988.

7. Murphy and Henderson 1997.

8. Johnston 1908:269, footnote 2; Greene and Wiseman 2023.

9. Bentley 1900:395.

10. Loveridge 1931.

11. Joy 1951.

12. Pruitt 1991. I have seen men with long and sharp thumbnails in Congo on a regular basis, including Aristote.

13. Branch 1984.

14. Besant 2013.

15. McLynn 1993.

16. Ibid.

17. Branch and Ha[a]cke 1980.

18. Greene and Wiseman 2023.

19. Because the Philippines were a Spanish colony for many years until 1898, many colonial-era names are still in use. *Negritos* means "little blacks," a problematic term similar to *pygmy* (Deaderick 2020).

20. Headland and Greene 2011.

21. To be clear, there is no direct evidence for this, but given the totality of the current and historical evidence, it seems likely.

22. Harcourt-Smith 2010; Head and Müller 2020; Roberts 2022.

23. Carroll 2016; Gibbons 2024.

24. Greene 1997; D. Hart and Sussman 2009.

25. Starin and Burghardt 1992.
26. Crude stone tools have been found at Lomekwi, Kenya, that are 3.3 million years old, but it is unclear whether hominins could have wielded them as weapons (Gibbons 2024).
27. Diamond 1987.
28. Britz et al. 2020.
29. Greenbaum and Kusamba 2012; Portillo and Greenbaum 2014; Kujirakwinja et al. 2019.
30. Greenbaum et al. 2023.
31. Mora et al. 2011; Engel et al. 2021.
32. Raxworthy and Tilston Smith 2021.
33. Space precludes me from discussing similar issues with Nagoya Protocol permits, which came into effect in 2014. This, too, is a well-intentioned set of regulations, but they are discouraging biodiversity surveys and collection of natural history specimens (e.g., Britz et al. 2020). At least one of my colleagues has been so exasperated by this new labyrinth of red tape that they have abandoned a decades-old research program in an area of the world that desperately needs one.
34. Rohwer et al. 2022.
35. Pennisi 2024.
36. Nietzel 2023.
37. Pennisi 2024.
38. Britz et al. 2020.
39. Engel et al. 2021:385.
40. Hope et al. 2018; Nachman et al. 2023.
41. Nachman et al. 2023.
42. M. Heinicke et al. 2023.
43. Pierce 2020.
44. Greenbaum 2017.
45. Timmons 2023.
46. Kolbert 2014. For a contrary viewpoint, see Wiens and Saban (2025).
47. Ceballos et al. 2017:E6095.
48. Ceballos and Erhlich 2023.
49. González-del-Pliego et al. 2019; Borgelt et al. 2022; Re:wild et al. 2023.
50. Greenbaum et al. 2015; Ghose et al. 2023.
51. Re:wild et al. 2023.
52. Meiri et al. 2023.
53. Sengupta and Friedman 2021.

54. An executive order on the very first day of the second Trump administration pulled the United States out of the 2015 Paris Agreement (Nilsen et al. 2025).

55. Dinerstein et al. 2019; Malhi et al. 2020; Vince 2022; Barnhart 2023.

56. Pecl et al. 2017; Bogitsch et al. 2019; Chason et al. 2023.

57. Duarte et al. 2020; The Cenozoic CO2 Proxy Integration Project (CenCO2 PIP) Consortium 2023; Milman 2024.

58. Jacobson et al. 2019; Duarte et al. 2020; Karnauskas et al. 2020; Pang et al. 2020; Jin et al. 2022.

59. Duarte et al. 2020; Karnauskas et al. 2020. In crowded indoor spaces, carbon dioxide levels can exceed 2,000 ppm (Pang et al. 2020).

60. Gates 2021.

61. Exceeding the threshold of 1.5 degrees Celsius for one year does not mean that we have breached the critical boundary that scientists have been warning about, at least not yet. The average is calculated over decades, but we are now flirting with disaster, because every tenth of a degree of warming makes the problem worse (N. Jones 2023; Irfan 2024).

62. Borunda 2025.

63. Cuff 2025.

64. Urban 2024.

65. Pecl et al. 2017; D. Wagner 2020.

66. Giam 2017; Kreier 2022.

67. Moat and Purdon 2021.

68. H. J. Hawkins et al. 2023.

69. Brinkmann et al. 2019.

70. C. Einhorn 2022.

71. N. Harris et al. 2021. Also consider that all of sub-Saharan Africa (excluding the relatively developed country of South Africa) is responsible for less than 1 percent of total global emissions (Wallace-Wells 2022).

72. Gatti et al. 2021.

73. Kreier 2022.

74. Kahlenberg et al. 2020.

75. Heinrich et al. 2022.

76. Johnson et al. 2017; Bello et al. 2024.

77. Sage 2020.

78. Ibid.:20.

79. Koch et al. 2019:20; Roberts 2022.

80. Koch et al. 2019.

81. Roberts 2022.

82. Roberts 2022; Anonymous 2024b.

83. Koch and Kaplan 2022.

84. Bastin et al. 2019.

85. C. Einhorn 2022.

86. Hall et al. 2023; Hood 2023. However, if the forests are fragmented, dispersal of seeds by some animal pollinators might be limited, thus slowing the restoration process (Bello et al. 2024).

87. Koch and Kaplan 2022; Hall et al. 2023:World Resources Institute map.

88. B. Williams et al. 2024.

89. Mitchell and Pleeck 2022.

90. Callahan and Mankin 2023; Anonymous 2024c; Wallace-Wells 2024.

91. Griscom et al. 2020.

92. Zanon 2020; Kaiser et al. 2021; Anonymous 2022; Hall et al. 2023; Haya et al. 2023.

93. Kahlenberg et al. 2020.

94. In February 2025, Aristote sent me a WhatsApp message to say that he had fled his home in Kavumu (about twenty miles north of Bukavu) with his family. They were temporarily safe from the rebels in a remote forest village, but they had no food, money, or medicine. During the fighting a bomb damaged his house.

95. Fuller et al. 2019; J. Graham 2022; Beardsley 2023; Peltier 2023; Walsh 2023; Mednick 2024; United Nations 2024; Alonga and Mureithi 2025; Rolley and Farge 2025.

96. Van Ittersum et al. 2016; G. Li et al. 2024.

97. The Kivu-Kinshasa Green Corridor I mentioned in the preface is a great start, but there are many other "undisturbed forests" of Congo that can and should receive protection (G. Einhorn and de Merode 2025).

98. Hood 2023.

99. Hood 2023; B. Jones 2024.

REFERENCES

Afelt, Aneta, Roger Frutos, and Christian Devaux. 2018. "Bats, Coronaviruses, and Deforestation: Toward the Emergence of Novel Infectious Diseases?" *Frontiers in Microbiology* 9: 702.

Agarwal, Ishan, Aaron M. Bauer, Varad B. Giri, and Akshay Khandekar. 2019. "An Expanded ND2 Phylogeny of the *brookii* and *prashadi* Groups with the Description of Three New Indian *Hemidactylus* Oken (Squamata: Gekkonidae)." *Zootaxa* 4619 (3): 431–58.

Agarwal, Ishan, Luis M. P. Ceríaco, Margarita Metallinou, Todd R. Jackman, and Aaron M. Bauer. 2021. "How the African House Gecko (*Hemidactylus mabouia*) Conquered the World." *Royal Society Open Science* 8 (8): 210749.

Ahossou, Oscar D., Kasso Daïnou, Steven B. Janssens, Ludwig Triests, and Olivier J. Hardy. 2020. "Species Delimitation and Phylogeography of African Tree Populations of the Genus *Parkia* (Fabaceae)." *Tree Genetics and Genomes* 16 (68).

Akani, Godrey C., Nwabueze Ebere, Daniel Franco, Edem A. Eniang, Fabio Petrozzi, Edoardo Politano, and Luca Luiselli. 2013. "Correlation Between Annual Activity Patterns of Venomous Snakes and Rural People in the Niger Delta, Southern Nigeria." *Journal of Venomous Animals and Toxins Including Tropical Diseases* 19: 2.

Alimohamadi, Yousef, Habteyes Hailu Tola, Abbas Abbasi-Ghahramanloo, Majid Janani, and Mojtaba Sepandi. 2021. "Case Fatality Rate of COVID-19: A Systematic Review and Meta-Analysis." *Journal of Preventive Medicine and Hygiene* 62 (2): E311–E320.

Allen, Kaitlin E., Walter P. Tapondjou N., Eli Greenbaum, Luke J. Welton, and Aaron M. Bauer. 2019. "High Levels of Hidden Phylogenetic Structure Within Central and West African *Trachylepis* Skinks." *Salamandra* 55 (4): 231–41.

Alonga, Ruth, and Carlos Mureithi. 2025. "'I Curse This War': Hunger and Fear in Goma After Rebel Takeover." *Guardian*, January 31, 2025. www.theguardian.com/world/2025/jan/31/hunger-fear-goma-m23-takeover-democratic-republic-of-the-congo.

Alsan, Marcella. 2015. "The Effect of the TseTse Fly on African Development." *American Economic Review* 105 (1): 382–410.

References

Alvar, Jorge, Margriet den Boer, and Daniel Argaw Dagne. 2021. "Towards the Elimination of Visceral Leishmaniasis as a Public Health Problem in East Africa: Reflections on an Enhanced Control Strategy and a Call for Action." *Lancet Global Health* 9: 1763–69.

Anderson, David L., Will Koomjian, Brian French, Scott R. Altenhoff, and James Luce. 2015. "Review of Rope-Based Access Methods for the Forest Canopy: Safe and Unsafe Practices in Published Information Sources and a Summary of Current Methods." *Methods in Ecology and Evolution* 6 (8): 865–72.

Anonymous. 1956. *Traveler's Guide to the Belgian Congo and the Ruanda Urundi.* 2nd ed. Brussels: Tourist Bureau for the Belgian Congo and Ruanda-Urundi.

Anonymous. 1964. "300 Flee Coquilhatville." *New York Times*, September 10, 1964. www.nytimes.com/1964/09/11/archives/300-flee-coquilhatville.html.

Anonymous. 2014. "Chito and Pocho." *NPR*, February 28, 2014. www.npr.org/2014/02/28/283934611/chito-and-pocho.

Anonymous. 2015. "Bookshelf: A Fresh Lens on the Notorious Episode of Ota Benga." *New York Times*, May 29, 2015. www.nytimes.com/2015/05/31/nyregion/a-fresh-lens-on-the-notorious-episode-of-ota-benga.html.

Anonymous. 2020. "Africa's Population Will Double by 2050." *Economist*, March 28, 2020. www.economist.com/special-report/2020/03/26/africas-population-will-double-by-2050.

Anonymous. 2022. "Costa Rica's Forest Conservation Pays Off." *World Bank Group*, November 16, 2022. www.worldbank.org/en/news/feature/2022/11/16/costa-rica-s-forest-conservation-pays-off.

Anonymous. 2023. "White Rhinos Reintroduced to DR Congo National Park." *Aljazeera*, June 11, 2023. www.aljazeera.com/news/2023/6/11/white-rhinos-reintroduced-to-dr-congo-national-park.

Anonymous. 2024a. "Products in US Supermarkets Linked to Deforestation of Tropical Forests." *global witness*, March 26, 2024. www.globalwitness.org/en/campaigns/forests/products-us-supermarkets-linked-deforestation-tropical-forests/.

Anonymous. 2024b. "Are We in the Anthropocene Yet?" *Nature (London)*, March 20, 2024. www.nature.com/articles/d41586-024-00815-0.

Anonymous. 2024c. "Global Warming Is Coming for Your Home." *Economist*, April 11, 2024. www.economist.com/leaders/2024/04/11/global-warming-is-coming-for-your-home.

Anonymous. 2025a. "Population, Total—Congo, Dem. Rep." *World Bank Group*, accessed February 2, 2025. https://data.worldbank.org/indicator/SP.POP.TOTL.

Anonymous. 2025b. "Which Everyday Products Contain Palm Oil?" WWF, accessed February 1, 2025. www.worldwildlife.org/pages/which-everyday-products-contain-palm-oil.

Anthes, Emily. 2022. "The Animal Translators." *New York Times*, August 30, 2022. www.nytimes.com/2022/08/30/science/translators-animals-naked-mole-rats.html.

Aronson, Naomi, Barbara L. Herwaldt, Michael Libman, Richard Pearson, Rogelio Lopez-Velez, Peter Weina, Edgar Carvalho, et al. 2017. "Guidelines: Diagnosis and Treatment of Leishmaniasis: Clinical Practice Guidelines by the Infectious Diseases Society of America (IDSA) and the American Society of Tropical Medicine and Hygiene (ASTMH)." *American Journal of Tropical Medicine and Hygiene* 96 (1): 24–45.

Arnold, E. N. 1989. "Systematics and Adaptive Radiation of Equatorial African Lizards Assigned to the Genera *Adolfus, Bedriagaia, Gastropholis, Holaspis* and *Lacerta* (Reptilia: Lacertidae)." *Journal of Natural History* 23 (3): 525–55.

Arnold, E. N. 2002. "*Holaspis*, a Lizard That Glided by Accident: Mosaics of Cooption and Adaptation in a Tropical Forest Lacertid (Reptilia, Lacertidae)." *Bulletin of the Natural History Museum of London (Zoology)* 68 (2): 155–63.

Badjedjea, Gabriel, Franck M. Masudi, Benjamin Dude Akaibe, and Václav Gvoždík. 2022. "Amphibians of Kokolopori: An Introduction to the Amphibian Fauna of the Central Congolian Lowland Forests, Democratic Republic of the Congo." *Amphibian and Reptile Conservation* 16 (1): 35–70.

Baisas, Laura. 2023. "Wild Bonobos Show Surprising Signs of Cooperation Between Groups." *Popular Science*, November 16, 2023. www.popsci.com/environment/bonobos-cooperations/.

Balandier, Georges. 1968. *Daily Life in the Kingdom of the Kongo: From the Sixteenth to the Eighteenth Century*. New York: Pantheon Books.

Ballard, Chris. 2006. "Strange Alliance: Pygmies in the Colonial Imaginary." *World Archaeology* 38 (1): 133–51.

Bampton, S. S. 1962. "The Institute for Scientific Research in Central Africa." *Nature (London)* 193: 130.

Barnes, Julian E. 2025. "C.I.A. Now Favors Lab Leak Theory to Explain Covid's Origins." *New York Times*, January 27, 2025. www.nytimes.com/2025/01/25/us/politics/cia-covid-lab-leak.html.

Barnhart, Max. 2023. "A 'Tropical Disease' Carried by Sand Flies Is Confirmed in a New Country: The U.S." *NPR*, November 1, 2023. www.npr.org/sections/goatsandsoda/2023/11/01/1209681147/leishmaniasis-sand-flies-tropical-disease-endemic-north-america-united-states.

References

Barthel, Matti, Marijn Bauters, Simon Baumgartner, Travis W. Drake, Nivens Mok-
wele Bey, Glenn Bush, Pascal Boeckx, et al. 2022. "Low N_2O and Variable CH_4
Fluxes from Tropical Forest Soils of the Congo Basin." *Nature Communications*
13: 330.

Bastin, Jean-François, Yelena Finegold, Claude Garcia, Danilo Mollicone, Marcelo
Rezende, Devin Routh, Constantin M. Zohner, et al. 2019. "The Global Tree
Restoration Potential." *Science (Washington)* 365 (6448): 76–79.

Bateman, P. W., and P. A. Fleming. 2009. "There Will Be Blood: Autohaemorrhage
Behaviour as Part of the Defence Repertoire of an Insect." *Journal of Zoology
(London)* 278 (4): 342–48.

Bates, Michael F., Krystal A. Tolley, Shelley Edwards, Zoë Davids, Jessica M. Da Silva,
and William R. Branch. 2013. "A Molecular Phylogeny of the African Plated
Lizards, Genus *Gerrhosaurus* Wiegmann, 1828 (Squamata: Gerrhosauridae),
with the Description of Two New Genera." *Zootaxa* 3750 (5): 465–93.

Bauer, Hans, Amy Dickman, Guillaume Chapron, Alayne Oriol-Cotterill, Samantha
K. Nicholson, Claudio Sillero-Zubiri, Luke Hunter, et al. 2022. "Threat
Analysis for More Effective Lion Conservation." *Oryx* 56 (1): 108–15.

Bauer, Hans, Guillaume Chapron, Kristin Nowell, Philipp Henscheld, Paul Fun-
ston, Luke T. B. Hunter, David W. Macdonald, et al. 2015. "Lion (*Panthera
leo*) Populations are Declining Rapidly Across Africa, Except in Intensively
Managed Areas." *Proceedings of the National Academy of Sciences (USA)* 112
(48): 14894–99.

Bawa, K. S. 1990. "Plant-Pollinator Interactions in Tropical Rain Forests." *Annual
Review of Ecology and Systematics* 21: 399–422.

Beardsley, Eleanor. 2023. "Recent Coups in Africa Have an Effect on At Least 1
Country in Europe: France." *NPR*, September 29, 2023. www.npr.org/2023/
09/29/1202582084/recent-coups-in-africa-have-an-effect-on-at-least-1-country-
in-europe-france.

Bébé Ngouateu, Omer, and Blaise Dondji. 2022. "Leishmaniasis in Cameroon and
Neighboring Countries: An Overview of Current Status and Control Challenges." *Current Research in Parasitology and Vector-Borne Diseases* 2: 100077.

Bédécarrats, Florent, Oriane Lafuente-Sampietro, Martin Leménager, and Dominique
Lukono Sowa. 2019. "Building Commons to Cope with Chaotic Urbaniza-
tion? Performance and Sustainability of Decentralized Water Services in the
Outskirts of Kinshasa." *Journal of Hydrology* 573: 1096–108.

Bello, Carolina, Thomas W. Crowther, Danielle Leal Ramos, Teresa Morán-López,
Marco A. Pizo, and Daisy H. Dent. 2024. "Frugivores Enhance Potential

Carbon Recovery in Fragmented Landscapes." *Nature Climate Change* 14: 636–43.

Benítez-López, Ana, Luca Santini, Aafke M. Schipper, Michela Busana, and Mark A. J. Huijbregts. 2019. "Intact but Empty Forests? Patterns of Hunting-Induced Mammal Defaunation in the Tropics." *PLoS Biology* 17: e3000247.

Bentley, W. Holman. 1900. *Pioneering on the Congo.* Vol. 1. New York: Fleming H. Revell.

Benton, Michael J., and James M. Clark. 1988. "Archosaur Phylogeny and the Relationships of the Crocodylia." In *The Phylogeny and Classification of the Tetrapods*, vol. 1: *Amphibians, Reptiles, Birds*, Systematics Association Special Volume No. 35A, edited by Michael J. Benton, 295–338. Oxford: Clarendon Press.

Beolens, Bo, Michael Watkins, and Michael Grayson. 2011. *The Eponym Dictionary of Reptiles.* Baltimore: Johns Hopkins University Press.

Besant, Alexander. 2013. "Zambia Man Bites Python That Tried to Kill Him." *GlobalPost*, May 26, 2013. https://theworld.org/stories/2013-05-26/zambia-man-bites-python-tried-kill-him.

Besenbacher, Søren, Christina Hvilsom, Tomas Marques-Bonet, Thomas Mailund, and Mikkel Heide Schierup. 2019. "Direct Estimation of Mutations in Great Apes Reconciles Phylogenetic Dating." *Nature Ecology and Evolution* 3: 286–92.

Bessone, Mattia, Hjalmar S. Kühl, Gottfried Hohmann, Ilka Herbinger, Kouame Paul N'Goran, Papy Asanzi, Pedro B. Da Costa, et al. 2020. "Drawn Out of the Shadows: Surveying Secretive Forest Species with Camera Trap Distance Sampling." *Journal of Applied Ecology* 57 (5): 963–74.

Bhatt, Deepak L., Renato D. Lopes, and Robert A. Harrington. 2022. "Diagnosis and Treatment of Acute Coronary Syndromes: A Review." *Journal of the American Medical Association* 327 (7): 662–75.

BirdLife International. 2017. "*Corythaeola cristata.*" The IUCN Red List of Threatened Species 2017: e.T22688425A111660258. Accessed February 1, 2025. www.iucnredlist.org/species/22688425/264142930.

Bittencourt-Silva, Gabriela B., David Langerman, and Krystal A. Tolley. 2020. "Why the Long Finger? Observation of Male–Male Combat in African Bush Squeaker Frogs, *Arthroleptis stenodactylus* (Anura: Arthroleptidae)." *Herpetological Bulletin* 151: 45.

Blackburn, David C. 2008. "Biogeography and Evolution of Body Size and Life History of African Frogs: Phylogeny of Squeakers (*Arthroleptis*) and Long-Fingered

Frogs (*Cardioglossa*) Estimated from Mitochondrial Data." *Molecular Phylogenetics and Evolution* 49 (3): 806–26.

Blackburn, David C. 2009. "Diversity and Evolution of Male Secondary Sexual Characters in African Squeakers and Long-Fingered Frogs." *Biological Journal of the Linnean Society* 96 (3): 553–73.

Blackburn, David C., Christian Boix, Eli Greenbaum, Marissa Fabrezi, Danny Meirte, Andrew J. Plumptre, and Edward L. Stanley. 2016. "The Distribution of the Bururi Long-fingered Frog (*Cardioglossa cyaneospila*, Family Arthroleptidae), a Poorly Known Albertine Rift Endemic." *Zootaxa* 4170 (2): 355–64.

Blackburn, David C., Stuart V. Nielsen, Sonia L. Ghose, Marius Burger, LeGrand Nono Gonwouo, Eli Greenbaum, Václav Gvoždík, et al. 2021. "Phylogeny of African Long-Fingered Frogs (Arthroleptidae: *Cardioglossa*) Reveals Recent Allopatric Divergences in Coloration." *Ichthyology and Herpetology* 109 (3): 728–42.

Blackburn, David C., Violette Dérozier, Apoluke Djaman, Eli Greenbaum, and Gregory F. M. Jongsma. 2022. "First Record of Spiny Frogs, *Acanthixalus* (Hyperoliidae, Kassininae), from South of the Congo River." *Herpetology Notes* 15: 641–47.

Bogitsch, Burton J., Clint E. Carter, and Thomas N. Oeltmann. 2019. *Human Parasitology*. 5th ed. London: Academic Press.

Böhme, Wolfgang. 2000. "Diversity of a Snake Community in a Guinean Rain Forest (Reptilia, Serpentes)." In *Isolated Vertebrate Communities in the Tropics*, edited by Goetz Rheinwald, 69–78. Proceedings of the 4th International Symposium in Bonn. *Bonner Zoologische Monographien* 46.

Bonin, Franck, Bernard Devaux, and Alain Dupré. 2006. *Turtles of the World*. Baltimore: Johns Hopkins University Press.

Borgelt, Jan, Martin Dorber, Marthe Alnes Høiberg, and Francesca Verones. 2022. "More Than Half of Data Deficient Species Predicted to Be Threatened by Extinction." *Communications Biology* 5: 679.

Borunda, Alejandra. 2025. "2024 Was the Hottest Year on Record. The Reason Remains a Science Mystery." *NPR*, January 10, 2024. www.npr.org/2025/01/10/nx-s1-5232139/2024-hottest-year-human-history-global-warming.

Bourgoin, C., G. Ceccherini, M. Girardello, C. Vancutsem, V. Avitabile, P. S. A. Beck, R. Beuchle, et al. 2024. "Human Degradation of Tropical Moist Forests is Greater Than Previously Estimated." *Nature (London)* 631: 570–76.

Bradford, Phillips V., and Harvey Blume. 1992. *Ota: The Pygmy in the Zoo*. New York: St. Martin's.

Branch, W. R. 1984. "Pythons and People: Predators and Prey." *African Wildlife* 38 (6): 236–41.

Branch, W. R., and W. D. Ha[a]cke. 1980. "A Fatal Attack on a Young Boy by an African Rock Python *Python sebae.*" *Journal of Herpetology* 14 (3): 305–7.

Brander, L. M., R. de Groot, J. P. Schägner, V. Guisado-Goñi, V. van 't Hoff, S. Solomonides, A. McVittie, et al. 2024. "Economic Values for Ecosystem Services: A Global Synthesis and Way Forward." *Ecosystem Services* 66: 101606.

Brecko, J. and O. S. G. Pauwels. 2024. "Lacertidae: *Holaspis guentheri* & *Holaspis laevis*: Ultraviolet (UV) Biofluorescence." *African Herp News* 86: 30–34.

Brenzel, Kathleen Norris. 2007. *Sunset Western Garden Book*. Menlo Park, CA: Sunset Publishing.

Brinkmann, Nicole, Dominik Schneider, Josephine Sahner, Johannes Ballauff, Nur Edy, Henry Barus, Bambang Irawan, et al. 2019. "Intensive Tropical Land Use Massively Shifts Soil Fungal Communities." *Scientific Reports* 9: 3403.

Britannica, The Editors of Encyclopaedia. 2007. "Matadi: Democratic Republic of the Congo." *Encyclopedia Britannica*, March 16, 2007. www.britannica.com/place/Matadi.

Britz, Ralf, Anna Hundsdörfer, and Uwe Fritz. 2020. "Funding, Training, Permits— The Three Big Challenges of Taxonomy." *Megataxa* 001 (1): 049–052.

Broadley, Donald G. 1998. "Introduction to Reptilia: Karl Patterson Schmidt and the Reptile Volumes Based on the Collections of the American Museum of Natural History Congo Expedition 1909–1915." In *Contributions to the Herpetology of the Belgian Congo*, by Karl P. Schmidt and G. K. Noble, ix–xxxiii. Ithaca, NY: Society for the Study of Amphibians and Reptiles.

Broadley, Donald G., Krystal A. Tolley, Werner Conradie, Sarah Wishart, Jean-François Trape, Marius Burger, Chifundera Kusamba, et al. 2018. "A Phylogeny and Genus-Level Revision of the African File Snakes *Gonionotophis* Boulenger (Squamata: Lamprophiidae)." *African Journal of Herpetology* 67 (1): 43–60.

Browne, Ed. 2021. "Rainforest Home to Elephants, Other Endangered Species Taken Off World Heritage in Danger List." *Newsweek*, July 22, 2021. www.newsweek.com/salonga-rainforest-home-elephants-bonobos-taken-off-danger-list-1612137.

Bryan, Susan M. 2023. "Dozens Gather to Watch Endangered Tortoise Release on New Mexico Ranch." *HuffPost*, September 23, 2023. www.huffpost.com/entry/bc-us-saving-tortoises_n_650f5828e4b088d5608c47db.

Buel, J. W. 1890. *Heroes of the Dark Continent*. Philadelphia: Historical Publishing.

Burbrink, Frank T., Felipe G. Grazziotin, R. Alexander Pyron, David Cundall, Steve Donnellan, Frances Irish, J. Scott Keogh, et al. 2020. "Interrogating Genomic-Scale Data for Squamata (Lizards, Snakes, and Amphisbaenians) Shows No

References

Support for Key Traditional Morphological Relationships." *Systematic Biology* 69 (3): 502–20.

Busby, Mattha. 2022. "Return of the Rhino: Can We Bring the Northern White Back from Extinction?" *Guardian*, December 18, 2022. www.theguardian.com/science/2022/dec/18/return-of-the-northern-white-rhino-back-from-extinction.

Butler, Richard J., Stephen L. Brusatte, Mike Reich, Sterling J. Nesbitt, Rainer R. Schoch, and Jahn J. Hornung. 2011. "The Sail-Backed Reptile *Ctenosauriscus* from the Latest Early Triassic of Germany and the Timing and Biogeography of the Early Archosaur Radiation." *PLoS ONE* 6 (10): e25693.

Callahan, Christopher W., and Justin S. Mankin. 2023. "Persistent Effect of El Niño on Global Economic Growth." *Science (Washington)* 380 (6649): 1064–69.

Candy, Mhorag. 1984. "Habits and Breeding Biology of the Great Blue Turaco *Corythaeola cristata.*" *Journal of the East Africa Natural History Society and National Museum* (190): 1–19.

Carroll, Sean B. 2016. *The Serengeti Rules: The Quest to Discover How Life Works and Why it Matters.* Princeton, NJ: Princeton University Press.

Casewell, Nicholas R., Timothy N. W. Jackson, Andreas H. Laustsen, and Kartik Sunagar. 2020. "Causes and Consequences of Snake Venom Variation." *Trends in Pharmacological Sciences* 41 (8): 570–81.

Cauret, Caroline M. S., Marie-Theres Gansauge, Andrew S. Tupper, Benjamin L. S. Furman, Martin Knytl, Xue-Ying Song, Eli Greenbaum, et al. 2020. "Developmental Systems Drift and the Drivers of Sex Chromosome Evolution." *Molecular Biology and Evolution* 37 (3): 799–810.

Ceballos, Gerardo, and Paul R. Ehrlich. 2023. "Mutilation of the Tree of Life via Mass Extinction of Animal Genera." *Proceedings of the National Academy of Sciences (USA)* 120 (39): e2306987120.

Ceballos, Gerardo, Paul R. Ehrlich, and Rodolfo Dirzo. 2017. "Biological Annihilation via the Ongoing Sixth Mass Extinction Signaled by Vertebrate Population Losses and Declines." *Proceedings of the National Academy of Sciences (USA)* 114 (30): E6089–E6096.

Cenozoic CO2 Proxy Integration Project (CenCO2 PIP) Consortium, The. 2023. "Toward a Cenozoic History of Atmospheric CO2." *Science (Washington)* 382 (6675): eadi5177.

Ceríaco, Luis M. P., Aaron M. Bauer, Chifundera Kusamba, Ishan Agarwal, and Eli Greenbaum. 2021. "A New Species of Ground-Dwelling *Hemidactylus* (Squamata: Gekkonidae) from Southwestern Democratic Republic of the Congo." *Journal of Herpetology* 55 (2): 105–11.

Ceríaco, Luis M. P., Sango dos Anjos Carlos de Sá, and Aaron M. Bauer. 2018. "The Genus *Osteolaemus* (Crocodylidae) in Angola and a New Southernmost Record for the Genus." *Herpetology Notes* 11: 337–41.

Chabiron, Clothilde, Silvia Gally, and Didier Demolin. 2013. "Les Parlers Pygmées du Bassin Équatorial du Congo." *Géolinguistique* 14: 125–44.

Chaney, Teslin, Olivier S. G. Pauwels, Zoltán Nagy, Václav Gvoždík, Chifundera Kusamba, Gabriel Badjedjea, Franck M. Masudi, et al. 2024. "Phylogenetics and Integrative Taxonomy of African Water Snakes (Squamata: Colubridae: *Grayia*)." *Herpetological Monographs* 38 (1): 1–52.

Chapin, James P. 1939. "The Birds of the Belgian Congo. Part II." *Bulletin of the American Museum of Natural History* 75: iii–vii, 3–632.

Charlesworth, Deborah and John H. Willis. 2009. "The Genetics of Inbreeding Depression." *Nature Reviews Genetics* 10: 783–96.

Chason, Rachel, Kevin Crowe, John Muyskens, and Jahi Chikwendiu. 2023. "Where Malaria Is Spreading." *Washington Post*, October 23, 2023. www.washingtonpost.com/climate-environment/interactive/2023/malaria-disease-spread-climate-change-warming/.

Chauhan, Virander S., Chetan E. Chitnis, and Deepak Gaur. 2017. "Introduction: An Overview of Malaria and *Plasmodium*." In *Advances in Malaria Research*, edited by Deepak Gaur, Chetan E. Chitnis, and Virander S. Chauhan, 1–7. Hoboken, NJ: John Wiley.

Chavan, Utkarsha, and Manoj R. Borkar. 2023. "Observations on Cooperative Fishing, Use of Bait for Hunting, Propensity for Marigold Flowers and Sentient Behaviour in Mugger Crocodiles *Crocodylus palustris* (Lesson, 1831) of River Savitri at Mahad, Maharashtra, India." *Journal of Threatened Taxa* 15 (8): 23750–62.

Chippaux, Jean-Philippe, and Kate Jackson. 2019. *Snakes of Central and Western Africa*. Baltimore: Johns Hopkins University Press.

Chirio, L., T. Diagne, and O. S. G. Pauwels. 2017. "*Cycloderma aubryi*." The IUCN Red List of Threatened Species 2017: e.T163448A1009393. Accessed March 22, 2024. https://dx.doi.org/10.2305/IUCN.UK.2017-3.RLTS.T163448A1009393.en.

Choudhury, Ananyo, Shaun Aron, Laura R. Botigué, Dhriti Sengupta, Gerrit Botha, Taoufik Bensellak, Gordon Wells, et al. 2020. "High-Depth African Genomes Inform Human Migration and Health." *Nature (London)* 586: 741–48.

Christy, Cuthbert. 1924. *Big Game and Pygmies: Experiences of a Naturalist in Central African Forests in Quest of the Okapi*. London: Macmillan.

References

Clement, Wendy L., Sam Bruun-Lund, Alanna Cohen, Finn Kjellberg, George D. Weiblen, and Nina Rønsted. 2020. "Evolution and Classification of Figs (*Ficus, Moraceae*) and Their Close Relatives (Castilleae) United by Involucral Bracts." *Botanical Journal of the Linnean Society* 193 (3): 316–39.

Coates, David J., Margaret Byrne, and Craig Moritz. 2018. "Genetic Diversity and Conservation Units: Dealing with the Species-Population Continuum in the Age of Genomics." *Frontiers in Ecology and Evolution* 6: 165.

Cohen, John. 2022. "From 'Open-Minded' to 'Underwhelming,' Mixed Reactions Greet Latest COVID-19 Origin Report." *Science (Washington)*, June 9, 2022. www.science.org/content/article/open-minded-underwhelming-mixed-reactions-greet-latest-covid-19-origin-report.

Collet, Marcel, and Jean-François Trape. 2020. "Une Nouvelle et Remarquable Espèce de *Naja* Semi-Aquatique (Elapidae, sous-genre *Boulengerina* Dollo, 1886) de la République Démocratique du Congo." *Bulletin de la Société Herpétologique de France* 173: 41–52.

Compere, Pierre, and Jean-Jacques Symoens. 1987. "Bassin du Zaïre/Zaïre Basin." In *African Wetlands and Shallow Water Bodies/Zones Humides et Lacs Peu Profonds d'Afrique, Directory Repertoire*, edited by M. J. Burgis and J. J. Symoens, 401–56. Paris: Éditions de l'ORSTOM.

Conrad, Joseph. 1980. *Heart of Darkness*. Norwalk, CT: Easton. First published in 1899.

Cordell, Dennis D., Janet MacGaffey, Wyatt MacGaffey, and James Oladipo Adejuwon. 2025. "Kinshasa." *Encyclopaedia Britannica*, February 18, 2025. www.britannica.com/place/Kinshasa.

Correa, Roberto J., Peter A. Lindsey, Rob Critchlow, Colin M. Beale, Jonas Geldmann, and Andrew J. Plumptre. 2024. "Performance of Protected Areas in Conserving African Elephants." *Conservation Letters* 2024: e13041.

Craft, Christopher. 2022. *Creating and Restoring Wetlands: From Theory to Practice*. 2nd ed. Amsterdam: Elsevier.

Craig, Jess. 2024. "Mpox Never Stopped Spreading in Africa. Now It's an International Public Health Emergency. Again." *Vox*, August 14, 2024. www.vox.com/future-perfect/366903/mpox-monkeypox-africa-continental-emergency-drc-who-clade.

Crawford, Dorothy H. 2018. *Viruses: A Very Short Introduction*. 2nd ed. Oxford: Oxford University Press.

Crezee, Bart, Greta C. Dargie, Corneille E. N. Ewango, Edward T. A. Mitchard, Ovide Emba B., Joseph Kanyama T., Pierre Bola, et al. 2022. "Mapping Peat Thickness and Carbon Stocks of the Central Congo Basin Using Field Data." *Nature Geoscience* 15: 639–44.

Crocodile Specialist Group. 1996. "*Osteolaemus tetraspis*." The IUCN Red List of Threatened Species 1996: e.T15635A4931429. Accessed February 19, 2025. https://dx.doi.org/10.2305/IUCN.UK.1996.RLTS.T15635A4931429.en.

Cuevas, Eduardo. 2024. "Mpox Strain Identified for First Time in US, Health Officials Say." *USA Today*, November 16, 2024. www.usatoday.com/story/news/health/2024/11/16/clade-i-mpox-strain-cdc/76363223007/.

Cuff, Madeleine. 2025. "Have We Already Breached the 1.5°C Global Warming Target?" *New Scientist*, February 10, 2025. www.newscientist.com/article/2467521-have-we-already-breached-the-1-5c-global-warming-target/.

Curran, C. H., and Carl Kauffeld. 1937. *Snakes and Their Ways*. New York: Harper.

Curry-Lindahl, Kai. 1974. "Conservation Problems and Progress in Equatorial African Countries." *Environmental Conservation* 1 (2): 111–22.

Cuthill, Innes C., Julian C. Partridge, Andrew T. D. Bennett, Stuart C. Church, Nathan S. Hart, and Sarah Hunt. 2000. "Ultraviolet Vision in Birds." *Advances in the Study of Behavior* 29: 159–214.

Daley, Jason. 2016. "Researchers 'Translate' Bat Talk. Turns Out, They Argue—A Lot." *Smithsonian Magazine*, December 23, 2016. www.smithsonianmag.com/smart-news/researchers-translate-bat-talk-and-they-argue-lot-180961564/.

Daniel, Ari. 2024. "Ebola Vaccine Cuts Death Rates in Half—Even if It's Given After Infection." *NPR*, February 15, 2024. www.npr.org/sections/goatsandsoda/2024/02/15/1231249465/ebola-vaccine-cuts-death-rates-in-half-even-if-its-given-after-infection.

Dargie, Greta C., Simon L. Lewis, Ian T. Lawson, Edward T. A. Mitchard, Susan E. Page, Yannick E. Bocko, and Suspense A. Ifo. 2017. "Age, Extent and Carbon Storage of the Central Congo Basin Peatland Complex." *Nature (London)* 542: 86–90.

Das, Sandeep, Saunak Pal, Sasidharan Siddharth, Muhamed Jafer Palot, Veerappan Deepak, and Surya Narayanan. 2022. "A New Species of Large-Bodied *Hemidactylus* Goldfuss, 1820 (Squamata: Gekkonidae) from the Western Ghats of India." *Vertebrate Zoology* 72: 81–94.

Davis, W. E. 1940. *Ten Years in the Congo*. New York: Reynal and Hitchcock.

Deaderick, Lisa. 2020. "Latina Professors Discuss Use of 'Negrito' and 'Negrita' in Latin Culture, After J.Lo Controversy." *San Diego Union-Tribune*, November 1, 2020. www.sandiegouniontribune.com/columnists/story/2020-11-01/latina-professors-discuss-use-of-negrito-negrita-in-latin-culture-after-j-lo-controversy.

References

Debonnet, Guy, and Kes Hillman-Smith. 2004. "Supporting Protected Areas in a Time of Political Turmoil: The Case of World Heritage Sites in the Democratic Republic of Congo." *Parks* 14 (1): 9–16.

De Manuel, Marc, Martin Kuhlwilm, Peter Frandsen, Vitor C. Sousa, Tariq Desai, Javier Prado-Martinez, Jessica Hernandez-Rodriguez, et al. 2016. "Chimpanzee Genomic Diversity Reveals Ancient Admixture with Bonobos." *Science (Washington)* 354 (6311): 477–81.

De Roo, Bas. 2017. "Taxation in the Congo Free State, an Exceptional Case? (1885–1908)." *Economic History of Developing Regions* 32 (2): 97–126.

Deufel, Alexandra, and David Cundall. 2003. "Feeding in *Atractaspis* (Serpentes: Atractaspididae): A Study in Conflicting Functional Constraints." *Zoology* 106 (1): 43–61.

Deutsch-Feldman, Molly, Jonathan B. Parr, Corinna Keeler, Nicholas F. Brazeau, Varun Goel, Michael Emch, Jessie K. Edwards, et al. 2021. "The Burden of Malaria in the Democratic Republic of the Congo." *Journal of Infectious Diseases* 223 (11): 1948–52.

Devlin, Larry. 2007. *Chief of Station, Congo.* New York: PublicAffairs.

de Waal, Frans B. M., and Frans Lanting (photographs). 1997. *Bonobo: The Forgotten Ape.* Berkeley: University of California Press.

de Witte, Gaston F. 1959. "Contribution à la Faune Herpétologique du Congo Belge: Description de Trois Serpents Nouveaux." *Revue de Zoologie et de Botanique Africaines* 60 (3–4): 348–51.

de Witte, Gaston F. 1965. "Les Caméléons de l'Afrique Centrale (République Démocratique du Congo, République du Rwanda et Royaume du Burundi)." *Annales Musée Royal de l'Afrique Centrale, Sciences Zoologiques* 142: 1–215 + pl. I–XII.

de Witte, Gaston F., and Raymond Laurent. 1943. "Contribution à la Systématique des Boiginae du Congo Belge (Rept.)." *Revue de Zoologie et de Botanique Africaines* 37 (1–2): 157–89.

Diagne, Tomas, Fidele Bandele Engalenzibo, and Mama Mbusa Kalisya. 2013. "Turtle and Tortoise Surveys in the Democratic Republic of the Congo." *Turtle Survival* 2013: 45–46.

Diamond, Jared M. 1987. "Justifiable Killing of Birds?" *Nature* (London) 330: 423.

Diamond, Jared. 1999. *Guns, Germs, and Steel: The Fates of Human Societies.* New York: Norton.

Díaz, Sandra, and Yadvinder Malhi. 2022. "Biodiversity: Concepts, Patterns, Trends, and Perspectives." *Annual Review of Environment and Resources* 47: 31–63.

Di Campo, Therese. 2017. "For Congo's Pygmies, Expulsion and Forest Clearance End a Way of Life." *Reuters*, Jan 12, 2017. www.reuters.com/article/

us-congo-pygmies-widerimage/for-congos-pygmies-expulsion-and-forest-clearance-end-a-way-of-life-idUSKBN14W2IQ.

Dickie, Emily A., Federica Giordani, Matthew K. Gould, Pascal Mäser, Christian Burri, Jeremy C. Mottram, Srinivasa P. S. Rao, et al. 2020. "New Drugs for Human African Trypanosomiasis: A Twenty First Century Success Story." *Tropical Medicine and Infectious Disease* 5 (1): 29.

Dinerstein, E., C. Vynne, E. Sala, A. R. Joshi, S. Fernando, T. E. Lovejoy, J. Mayorga, et al. 2019. "A Global Deal for Nature: Guiding Principles, Milestones, and Targets." *Science Advances* 5: eaaw2869.

Diogo, Rui, Julia L. Molnar, and Barnard Wood. 2017. "Bonobo Anatomy Reveals Stasis and Mosaicism in Chimpanzee Evolution, and Supports Bonobos as the Most Appropriate Extant Model for the Common Ancestor of Chimpanzees and Humans." *Scientific Reports* 7: 608.

Dobiey, Maik, and Gernot Vogel. 2007. *Venomous Snakes of Africa/Giftschlangen Afrikas*. Frankfurt: Edition Chimaira.

Dowell, Stephanie A., Daniel M. Portik, Vivian de Buffrénil, Ivan Ineich, Eli Greenbaum, Sergios-Orestis Kolokotronis, and Evon R. Hekkala. 2016. "Molecular Data from Contemporary and Historical Collections Reveal a Complex Story of Cryptic Diversification in the *Varanus* (*Polydaedalus*) *niloticus* Species Group." *Molecular Phylogenetics and Evolution* 94 (Part B): 591–604.

Duarte, Carlos M., Lukasz Jaremko, and Mariusz Jaremko. 2020. "Hypothesis: Potentially Systemic Impacts of Elevated CO2 on the Human Proteome and Health." *Frontiers in Public Health* 8: 543322.

du Chaillu, Paul B. 1861. *Explorations and Adventures in Equatorial Africa; with Accounts of the Manners and Customs of the People, and of the Chase of the Gorilla, the Crocodile, Leopard, Elephant, Hippopotamus, and Other Animals*. New York: Harper.

Dyck, Jan. 1992. "Reflectance Spectra of Plumage Areas Colored by Green Feather Pigments." *Auk* 109 (2): 293–301.

Eaton, Mitchell J., Andrew Martin, John Thorbjarnarson, and George Amato. 2009. "Species-Level Diversification of African Dwarf Crocodiles (Genus *Osteolaemus*): A Geographic and Phylogenetic Perspective." *Molecular Phylogenetics and Evolution* 50 (3): 496–506.

Eban, Katherine, and Jeff Kao. 2022. "COVID-19 Origins: Investigating a 'Complex and Grave Situation' Inside a Wuhan Lab." *Vanity Fair*, October 28, 2022. www.vanityfair.com/news/2022/10/covid-origins-investigation-wuhan-lab.

Edwards, Phil. 2016. "All 512 Animals Teddy Roosevelt and His Son Killed on Safari." *Vox*, February 3, 2016. www.vox.com/2015/7/29/9067587/theodore-roosevelt-safari.

References

Ehret, Christopher. 2002. *The Civilizations of Africa: A History to 1800*. Charlottesville: University of Virginia Press.

Einhorn, Catrin. 2022. "Tree Planting Is Booming. Here's How That Could Help, Or Harm, the Planet." *New York Times*, March 14, 2022. www.nytimes.com/2022/03/14/climate/tree-planting-reforestation-climate.html.

Einhorn, Gill, and Emmanuel de Merode. 2025. "The Democratic Republic of Congo to Create the Earth's Largest Protected Tropical Reserve." World Economic Forum, January 22, 2025. www.weforum.org/stories/2025/01/congo-kivu-kinshasa-green-corridor/.

Emanuel, Gabrielle. 2024. "Why the Mpox Outbreak in the Democratic Republic of Congo Is Worrying Disease Docs." *NPR*. March 27, 2024. www.npr.org/sections/goatsandsoda/2024/03/27/1239276957/mpox-outbreak-democratic-republic-of-congo-deadlier-strain.

Emerson, Barbara. 1979. *Lepold II of the Belgians: King of Colonialism*. New York: St. Martin's Press.

Emina, Jacques B. O., Henry V. Doctor, and Yazoumé Yé. 2021. "Profiling Malaria Infection Among Under-Five Children in the Democratic Republic of Congo." *PLoS ONE* 16 (5): e0250550.

Emslie, R. 2020a. "*Diceros bicornis*." The IUCN Red List of Threatened Species 2020: e.T6557A152728945. Accessed March 23, 2022. https://dx.doi.org/10.2305/IUCN.UK.2020-1.RLTS.T6557A152728945.en.

Emslie, R. 2020b. "*Ceratotherium simum*." The IUCN Red List of Threatened Species 2020: e.T4185A45813880. Accessed March 24, 2022. https://dx.doi.org/10.2305/IUCN.UK.2020-1.RLTS.T4185A45813880.en.

Engel, Michael S., Luis M. P. Ceríaco, Gimo M. Daniel, Pablo M. Dellapé, Ivan Löbl, Milen Marinov, Roberto E. Reis, et al. 2021. "The Taxonomic Impediment: A Shortage of Taxonomists, Not the Lack of Technical Approaches." *Zoological Journal of the Linnean Society* 193 (2): 381–87.

Engelbrecht, Hanlie M., William R. Branch, Eli Greenbaum, Graham J. Alexander, Kate Jackson, Marius Burger, Werner Conradie, et al. 2019. "Diversifying into the Branches: Species Boundaries in African Green and Bush Snakes, *Philothamnus* (Serpentes: Colubridae)." *Molecular Phylogenetics and Evolution* 130: 357–65.

Evans, Ben J., Marie-Theres Gansauge, Edward L. Stanley, Benjamin L. S. Furman, Caroline M. S. Cauret, Caleb Ofori-Boateng, Václav Gvoždík, et al. 2019. "*Xenopus fraseri*: Mr. Fraser, Where Did Your Frog Come From?" *PLoS ONE* 14 (9): e0220892.

Evans, Ben J., Timothy F. Carter, Eli Greenbaum, Václav Gvoždík, Darcy B. Kelley, Patrick J. McLaughlin, Olivier S. G. Pauwels, et al. 2015. "Genetics, Morphology, Advertisement Calls, and Historical Records Distinguish Six New Polyploid Species of African Clawed Frog (*Xenopus*, Pipidae) from West and Central Africa." *PLoS ONE* 10 (2): e0142823.

Evans, Ben J., Václav Gvoždík, Martin Knytl, Caroline M. S. Cauret, Anthony Herrel, Eli Greenbaum, Jay Patel, et al. 2024. "Rapid Sex Chromosome Turnover in African Clawed Frogs (*Xenopus*) and the Origins of New Sex Chromosomes." *Molecular Biology and Evolution* 41 (12): msae234.

Fa, Julia E., Jesús Olivero, Miguel Angel Farfán, Jerome Lewis, Hirokazu Yasuoka, Andrew Noss, Shiho Hattori, et al. 2016. "Differences Between Pygmy and Non-Pygmy Hunting in Congo Basin Forests." *PLoS ONE* 11 (9): e0161703.

Faaborg, John. 2020. *Book of Birds: Introduction to Ornithology*. College Station: Texas A&M University Press.

Fairlamb, Alan H., and David Horn. 2018. "Melarsoprol Resistance in African Trypanosomiasis." *Trends in Parasitology* 34 (6): 481–92.

Fakhri, Nargis. 2018. "Why This Picture of the Last Male Northern White Rhino Was One of the Most Important Photos of 2018." *Time*, December 20, 2018. https://time.com/5482842/time-top-10-photos-2018-sudan-northern-white-rhino/.

Falisse, Jean-Benoît, and Alain Mpanya. 2022. "Clinical Trials as Disease Control? The Political Economy of Sleeping Sickness in the Democratic Republic of the Congo (1996–2016)." *Social Science and Medicine* 299: 114882.

Farmer, Paul. 2021. *Fevers, Feuds, and Diamonds: Ebola and the Ravages of History*. New York: Picador.

Farooq, Harith, Josué A. R. Azevedo, Amadeu Soares, Alexandre Antonelli, and Søren Faurby. 2021. "Mapping Africa's Biodiversity: More of the Same Is Just Not Good Enough." *Systematic Biology* 70 (3): 623–33.

Ferrari, Giovanfrancesco, Henry M. Ntuku, Sandro Schmidlin, Eric Diboulo, Antoinette K. Tshefu, and Christian Lengeler. 2016. "A Malaria Risk Map of Kinshasa, Democratic Republic of Congo." *Malaria Journal* 15: 27.

Figueroa, Alex, Alexander D. McKelvy, L. Lee Grismer, Charles D. Bell, and Simon P. Lailvaux. 2016. "A Species-Level Phylogeny of Extant Snakes with Description of a New Colubrid Subfamily and Genus." *PLoS ONE* 11 (9): e0161070.

Fonteyn, Davy, Cédric Vermeulen, Anaïs-Pasiphaé Gorel, Pedro Luiz Silva de Miranda, Simon Lhoest, and Adeline Fayolle. 2023. "Biogeography of Central

African Forests: Determinants, Ongoing Threats and Conservation Priorities of Mammal Assemblages." *Diversity and Distributions* 29 (6): 698–712.

Foster, Damon, and Craig Foster (producers and directors). 2013. *The Man Who Swims with Crocodiles*. Nat Geo Wildlife Documentary, Foster Brothers Film Productions.

Frederick, Adolphus (Duke of Mecklenburg). 1910. *In the Heart of Africa*. London: Cassell.

Fredriksson, Gabriella M. 2005. "Predation on Sun Bears by Reticulated Python in East Kalimantan, Indonesian Borneo." *Raffles Bulletin of Zoology* 53 (1): 165–68.

Freytas-Tamura, Kimiko de. 2018. "He's Handing Over the Presidency but Not Necessarily His Power." *New York Times*, December 14, 2018. www.nytimes.com/2018/12/14/world/africa/kabila-congo-elections.html.

Freytas-Tamura, Kimiko de. 2019. "After Tarnished Election, Opposition Figure Becomes Congo's President." *New York Times*, January 24, 2019. www.nytimes.com/2019/01/24/world/africa/congo-president-inauguration-tshisekedi-kabila.html.

Fromont, Cécile. 2023. "The Kongo Kingdom." In *Great Kingdoms of Africa*, edited by John Parker, 143–67. Oakland: University of California Press.

Frost, Darrel R. 2025. "Amphibian Species of the World: An Online Reference." Version 6.2. Accessed February 15, 2025. Electronic Database accessible at https://amphibiansoftheworld.amnh.org/index.php. American Museum of Natural History, New York. doi.org/10.5531/db.vz.0001.

Fruth, B., J. R. Hickey, C. André, T. Furuichi, J. Hart, T. Hart, H. Kuehl, et al. 2016. "*Pan paniscus* (errata version published in 2016)." The IUCN Red List of Threatened Species 2016: e.T15932A102331567. Accessed February 2, 2025. https://dx.doi.org/10.2305/IUCN.UK.2016-2.RLTS.T15932A17964305.en.

Fuller, Trevon L., Thomas P. Narins, Janet Nackoney, Timothy C. Bonebrake, Paul Seink Clee, Katy Morgan, Anthony Tróchez, et al. 2019. "Assessing the Impact of China's Timber Industry on Congo Basin Land Usage." *Area* 51 (2): 340–49.

Furuichi, Takeshi, and Jo Thompson, eds. 2008. *The Bonobos: Behavior, Ecology, and Conservation*. New York: Springer.

Galloway, Kent. 2021. *Courage in the Congo: A Doctor's Fight to Save the Pygmies*. Published by the author.

Gates, Bill. 2021. *How to Avoid a Climate Disaster: The Solutions We Have and the Breakthroughs We Need*. New York: Knopf.

Gatti, Luciana V., Luana S. Basso, John B. Miller, Manuel Gloor, Lucas Gatti Domingues, Henrique L. G. Cassol, Graciela Tejada, et al. 2021. "Amazonia as a Carbon Source Linked to Deforestation and Climate Change." *Nature (London)* 595: 388–93.

Ghazoul, Jaboury, and Douglas Sheil. 2010. *Tropical Rain Forest Ecology, Diversity, and Conservation.* Oxford: Oxford University Press.

Ghose, Sonia L., Tiffany A. Yap, Allison Q. Byrne, Hasan Sulaeman, Erica Bree Rosenblum, Alan Chan-Alvarado, Shruti Chaukulkar, et al. 2023. "Continent-Wide Recent Emergence of a Global Pathogen in African Amphibians." *Frontiers in Conservation Science* 4: 1069490.

Giam, Xingli. 2017. "Global Biodiversity Loss from Tropical Deforestation." *Proceedings of the National Academy of Sciences (USA)* 114 (23): 5775–77.

Gibbons, Ann. 2024. "Lucy's World." *Science (Washington)* 384 (6691): 20–25.

González-del-Pliego, Pamela, Robert P. Freckleton, David P. Edwards, Michelle S. Koo, Brett R. Scheffers, R. Alexander Pyron, and Walter Jetz. 2019. "Phylogenetic and Trait-Based Prediction of Extinction Risk for Data-Deficient Amphibians." *Current Biology* 29: 1557–63.

Gore, Meredith L., Robert Mwinyihali, Luc Mayet, Gavinet Duclair Makaya Baku-Bumb, Christian Plowman, and Michelle Wieland. 2021. "Typologies of Urban Wildlife Traffickers and Sellers." *Global Ecology and Conservation* 27: e01557.

Gorman, James. 2024. "Bats: A Love Story." *National Geographic* 246: 16–49.

Gossé, Koffi Jules, Sery Gonedelé-Bi, Sylvain Dufour, Emmanuel Danquah, and Philippe Gaubert. 2024. "Conservation Genetics of the White-Bellied Pangolin in West Africa: A Story of Lineage Admixture, Declining Demography, and Wide Sourcing by Urban Bushmeat Markets." *Ecology and Evolution* 14 (3): e11031.

Goyanes, Cristina. 2017. "These Badass Women Are Taking on Poachers—And Winning." *National Geographic*, October 12, 2017. www.nationalgeographic.com/adventure/article/black-mambas-anti-poaching-wildlife-rhino-team.

Graham, Alistair, and Peter Beard. 1990. *Eyelids of Morning: The Mingled Destinies of Crocodiles and Men.* San Francisco: Chronicle Books.

Graham, Jack. 2022. "The Next Amazon? Congo Basin Faces Rising Deforestation Threat." *Reuters*, November 11, 2022. www.reuters.com/business/cop/next-amazon-congo-basin-faces-rising-deforestation-threat-2022-11-11/.

Gramentz, Dieter. 2008. *African Flapshell Turtles*: Cyclanorbis *and* Cycloderma. Frankfurt: Edition Chimaira.

References

Greenbaum, Eli. 2017. *Emerald Labyrinth: A Scientist's Adventures in the Jungles of the Congo*. Lebanon, NH: ForeEdge. Copyright noted as 2018.

Greenbaum, Eli, and Chifundera Kusamba. 2012. "Conservation Implications Following the Rediscovery of Four Frog Species from the Itombwe Natural Reserve, Eastern Democratic Republic of the Congo." *Herpetological Review* 43 (2): 253–59.

Greenbaum, Eli, Daniel M. Portik, Kaitlin E. Allen, Eugene R. Vaughan, Gabriel Badjedjea, Michael F. Barej, Mathias Behangana, et al. 2022. "Systematics of the Central African Spiny Reed Frog *Afrixalus laevis* (Anura: Hyperoliidae), With the Description of Two New Species from the Albertine Rift." *Zootaxa* 5174 (3): 201–32.

Greenbaum, Eli, Jennifer Meece, Kurt D. Reed, and Chifundera Kusamba. 2015. "Extensive Occurrence of the Amphibian Chytrid Fungus in the Albertine Rift, a Central African Amphibian Hotspot." *Herpetological Journal* 25: 91–100.

Greenbaum, Eli, Kaitlin E. Allen, Eugene R. Vaughan, Olivier S. G. Pauwels, Van Wallach, Chifundera Kusamba, Wandege M. Muninga, et al. 2021. "Night Stalkers from Above: A Monograph of *Toxicodryas* Tree Snakes (Squamata: Colubridae) with Descriptions of Two New Cryptic Species from Central Africa." *Zootaxa* 4965 (1): 001–044.

Greenbaum, Eli, and Mingna V. Zhuang. 2019. "Robert Gravem Webb (1927–2018): Specialist on Trionychid Turtles and Mexican Herpetology." *Herpetological Review* 50 (3): 656–61.

Greenbaum, Eli, Nadezhda Galeva, and Michael Jorgensen. 2003. "Venom Variation and Chemoreception of the Viperid *Agkistrodon contortrix*: Evidence for Adaptation?" *Journal of Chemical Ecology* 29: 1741–55.

Greenbaum, Eli, Nancy Conkey, Chifundera Kusamba, Jennifer B. Pramuk, John L. Carr, Mark-Oliver Rödel, Kate Jackson, et al. 2012. "Systematics of Congo Basin True Toads (Anura: Bufonidae: *Amietophrynus*) Reveals Widespread Cryptic Speciation." Seventh World Congress of Herpetology abstract (no. 1411), Vancouver.

Greenbaum, Eli, Olivier S. G. Pauwels, Václav Gvoždík, Eugene R. Vaughan, Teslin Chaney, Michael Buontempo, Mwenebatu M. Aristote, et al. 2023. "Systematics of the Thirteen-Scaled Green Snake *Philothamnus carinatus* (Squamata: Colubridae) with the Description of a Cryptic New Species from Central and East Africa." *African Journal of Herpetology* 72 (2): 119–44.

Greenbaum, Eli, Samantha Stewart, Angelica Casas, Chifundera Kusamba, Wandege M. Muninga, Mwenebatu M. Aristote, and Javier Lobón-Rovira. 2025.

"Geographical Distribution. Gekkonidae. *Hemidactylus pfindaensis.*" *African Herp News* (87): 31–34.

Greene, Harry W. 1997. *Snakes: The Evolution of Mystery in Nature.* Berkeley: University of California Press.

Greene, Harry W., and Kevin D. Wiseman. 2023. "Heavy, Bulky, or Both: What Does 'Large Prey' Mean to Snakes?" *Journal of Herpetology* 57 (3): 340–66.

Greenwood, Steve (director and producer). 2007. *Nova: The Last Great Ape.* WGBH Educational Foundation for Public Broadcasting Service.

Gretener, Jessie. 2024. "World First IVF Rhino Pregnancy Could Save Northern White Rhinos from Extinction, Scientists Say." *CNN*, January 25, 2024. www.cnn.com/2024/01/25/africa/northern-white-rhino-ivf-scli-intl-scn/index.html.

Gribaldo, Simonetta, and Céline Brochier-Armanet. 2019. "Evolutionary Relationships Between Archaea and Eukaryotes." *Nature Ecology and Evolution* 4: 20–21.

Griscom, Bronson W., Jonah Busch, Susan C. Cook-Patton, Peter W. Ellis, Jason Funk, Sara M. Leavitt, Guy Lomax, et al. 2020. "National Mitigation Potential from Natural Climate Solutions in the Tropics." *Philosophical Transactions of the Royal Society B: Biological Sciences* 375 (1794): 20190126.

Groom, A. 2012. "*Parkia bicolor.*" The IUCN Red List of Threatened Species 2012: e.T19892546A20126291. Accessed February 19, 2025. https://dx.doi.org/10.2305/IUCN.UK.2012.RLTS.T19892546A20126291.en.

Grossman, Daniel. 2019. "Inside the Search for Africa's Carbon Time Bomb." *National Geographic*, October 1, 2019. www.nationalgeographic.com/science/article/inside-search-for-africa-carbon-time-bomb-peatland.

Groves, Colin P. 2013. "Tribe Gorillini: Gorillas." In *Mammals of Africa*, vol. 2: *Primates*, edited by Thomas M. Butynski, Jonathan Kingdon, and Jan Kalina, 35–38. London: Bloomsbury.

Guglielmi, Giorgia. 2024. "First Fossil Chromosomes Discovered in Freeze-Dried Mammoth Skin." *Nature (London)*, July 11, 2024. www.nature.com/articles/d41586-024-02253-4.

Hall, Jefferson S., Katherine Sinacore, and Michiel van Breugel. 2023. "Paying People to Replant Tropical Forests—And Letting Them Harvest the Timber—Can Pay Off for Climate, Justice and Environment." *Conversation*, December 15, 2023. https://theconversation.com/paying-people-to-replant-tropical-forests-and-letting-them-harvest-the-timber-can-pay-off-for-climate-justice-and-environment-219894.

Hallet, Jean-Pierre. 1967a. *Congo Kitabu.* Greenwich, CT: Fawcett.

References

Hallet, Jean-Pierre (with Alex Pelle). 1967b. *Animal Kitabu*. New York: Random House.

Ham, Anthony. 2023. "The Quest for a Crocodile Dictionary." *New York Times*, August 24, 2023. www.nytimes.com/2023/08/24/science/crocodile-dictionary-vocalizations.html.

Happold, Meredith. 2013. "Family Pteropodidae: Fruit Bats (Old World Fruit Bats)." In *Mammals of Africa*, vol. 4: *Hedgehogs, Shrews and Bats*, edited by Meredith Happold and David C. D. Happold, 223–27. London: Bloomsbury.

Happold, Meredith, and David C. D. Happold, eds. 2013. *Mammals of Africa*. Vol. 4: *Hedgehogs, Shrews and Bats*. London: Bloomsbury.

Harcourt-Smith, William H. E. 2010. "The First Hominins and the Origins of Bipedalism." *Evolution: Education and Outreach* 3: 333–40.

Harley, Eric H., Margaretha de Waal, Shane Murray, and Colleen O'Ryan. 2016. "Comparison of Whole Mitochondrial Genome Sequences of Northern and Southern White Rhinoceroses (*Ceratotherium simum*): The Conservation Consequences of Species Definitions." *Conservation Genetics* 17: 1285–91.

Harris, Nancy L., David A. Gibbs, Alessandro Baccini, Richard A. Birdsey, Sytze de Bruin, Mary Farina, Lola Fatoyinbo, et al. 2021. "Global Maps of Twenty-First Century Forest Carbon Fluxes." *Nature Climate Change* 11: 234–40.

Harris, Richard J., K. Anne-Isola Nekaris, and Bryan G. Fry. 2021. "Monkeying Around with Venom: An Increased Resistance to α-Neurotoxins Supports an Evolutionary Arms Race Between Afro-Asian Primates and Sympatric Cobras." *BMC Biology* 19: 253.

Hart, Donna, and Robert W. Sussman. 2009. *Man the Hunted: Primates, Predators, and Human Evolution*. Exp. ed. Philadelphia: Westview Press.

Hart, John A., and Jefferson S. Hall. 1996. "Status of Eastern Zaire's Forest Parks and Reserves." *Conservation Biology* 10 (2): 316–27.

Hawkins, Heidi-Jayne, Rachael I. M. Cargill, Michael E. Van Nuland, Stephen C. Hagen, Katie J. Field, Merlin Sheldrake, Nadeida A. Soudzilovskaia, et al. 2023. "Mycorrhizal Mycelium as a Global Carbon Pool." *Current Biology* 33: R560–R573.

Hawkins, John A., Maria E. Kaczmarek, Marcel A. Müller, Christian Drosten, William H. Press, and Sara L. Sawyer. 2019. "A Metaanalysis of Bat Phylogenetics and Positive Selection Based on Genomes and Transcriptomes from 18 Species." *Proceedings of the National Academy of Sciences (USA)* 116 (23): 11351–60.

Haya, Barbara K., Samuel Evans, Letty Brown, Jacob Bukoski, Van Butsic, Bodie Cabiyo, Rory Jaco, et al. 2023. "Comprehensive Review of Carbon

Quantification by Improved Forest Management Offset Protocols." *Frontiers in Forests and Global Change* 6: 958879.

Head, Jason J., and Johannes Müller. 2020. "Squamate Reptiles from Kanapoi: Faunal Evidence for Hominin Paleoenvironments." *Journal of Human Evolution* 140: 102451.

Headland, Thomas N., and Harry W. Greene. 2011. "Hunter-Gatherers and Other Primates as Prey, Predators, and Competitors of Snakes." *Proceedings of the National Academy of Sciences (USA)* 108 (52): E1470–E1474.

Heinicke, Matthew P., Stuart V. Nielsen, Aaron M. Bauer, Ryan Kelly, Anthony J. Geneva, Juan D. Daza, Shannon E. Keating, et al. 2023. "Reappraising the Evolutionary History of the Largest Known Gecko, the Presumably Extinct *Hoplodactylus delcourti*, via High-Throughput Sequencing of Archival DNA." *Scientific Reports* 13: 9141.

Heinicke, Stefanie, Roger Mundry, Christophe Boesch, Bala Amarasekaran, Abdulai Barrie, Terry Brncic, David Brugière, et al. 2019. "Advancing Conservation Planning for Western Chimpanzees Using IUCN SSC A.P.E.S.—The Case of a Taxon-Specific Database." *Environmental Research Letters* 14: 064001.

Heinrich, Viola H. A., Christelle Vancutsem, Ricardo Dalagnol, Thais M. Rosan, Dominic Fawcett, Celso H. L. Silva-Junior, Henrique L. G. Cassol, et al. 2022. "The Carbon Sink of Secondary and Degraded Humid Tropical Forests." *Nature (London)* 615: 436–42.

Hekkala, Evon, Matthew H. Shirley, George Amato, James D. Austin, Suellen Charter, John Thorbjarnarson, Kent A. Vliet, et al. 2011. "An Ancient Icon Reveals New Mysteries: Mummy DNA Resurrects a Cryptic Species Within the Nile Crocodile." *Molecular Ecology* 20 (20): 4199–215.

Hemingway, Ernest. 2005. *Under Kilimanjaro*. Edited by Robert W. Lewis and Robert E. Fleming. Kent, OH: Kent State University Press.

Hemingway, Ernest. 2015. *Green Hills of Africa*. The Hemingway Library Edition. New York: Scribner. First published in 1935.

Henderson, Donald M., and Robert Nicholls. 2015. "Balance and Strength—Estimating the Maximum Prey-Lifting Potential of the Large Predatory Dinosaur *Carcharodontosaurus saharicus*." *Anatomical Record* 298 (8): 1367–75.

Hernandez, Jessica. 2022. *Fresh Banana Leaves: Healing Indigenous Landscapes Through Indigenous Science*. Berkeley, CA: North Atlantic Books.

Hewlett, Barry S., ed. 2017. *Hunter-Gatherers of the Congo Basin: Cultures, Histories, and Biology of African Pygmies*. London: Routledge.

Heymann, David. 2014. "The Ebola Outbreak This Year That You Won't Have Heard About." *Guardian*, October 15, 2014. www.theguardian.com/commentisfree/2014/oct/15/ebola-outbreak-central-africa-contained.

References

Hildebrandt, Thomas B., Robert Hermes, Silvia Colleoni, Sebastian Diecke, Susanne Holtze, Marilyn B. Renfree, Jan Stejskal, et al. 2018. "Embryos and Embryonic Stem Cells from the White Rhinoceros." *Nature Communications* 9: 2589.

Hillman Smith, Kes. 2014. "The Northern White Rhinos." In *Garamba: Conservation in Peace and War*, edited by Kes Hillman Smith and José Kalpers, with Luis Arranz and Nuria Ortega, 244–70. Johannesburg, South Africa: Hirt and Carter.

Hime, Paul M., Alan R. Lemmon, Emily C. Moriarty Lemmon, Elizabeth Prendini, Jeremy M. Brown, Robert C. Thomson, Justin D. Kratovil, et al. 2021. "Phylogenomics Reveals Ancient Gene Tree Discordance in the Amphibian Tree of Life." *Systematic Biology* 701 (1): 49–66.

Hirschfeld, Mareike, David C. Blackburn, Marius Burger, Eli Greenbaum, Ange-Ghislain Zassi-Boulou, and Mark-Oliver Rödel. 2015. "Two New Species of Long-Fingered Frogs of the Genus *Cardioglossa* (Anura: Arthroleptidae) from Central African Rainforests." *African Journal of Herpetology* 64 (2): 81–102.

Ho, Vivian. 2025. "'Unknown Disease' That Can Kill Within Days Leaves 53 Dead in Congo." *Washington Post*, February 25, 2025. www.washingtonpost.com/world/2025/02/25/unknown-illness-hemorrhagic-fever-congo-africa/.

Hochschild, Adam. 1999. *King Leopold's Ghost: A Story of Greed, Terror, and Heroism in Colonial Africa*. New York: Houghton Mifflin.

Homer. 1712. *The Iliad of Homer, With Notes. To Which Are Prefix'd, a Large Preface, and the Life of Homer, by Madam Dacier. Done From the French by Mr. Ozell; and by Him Compar'd with the Greek. To Which Will be Made Some Farther Notes, That Shall be Added at the End of the Whole; by Mr. Johnson, late of Eton, Now of Bromford*. London: G. James, for Bernard Lintott, at the Cross-Keys Between the Two Temple-Gates.

Hood, Marlowe. 2023. "Football Pitch of Tropical Forest Lost Every 5 Seconds." Phys. Org, June 27, 2023. https://phys.org/news/2023-06-football-pitch-tropical-forest-lost.html.

Hope, Andrew G., Brett K. Sandercock, and Jason L. Malaney. 2018. "Collection of Scientific Specimens: Benefits for Biodiversity Sciences and Limited Impacts on Communities of Small Mammals." *BioScience* 68 (1): 35–42.

Hopkins, Helen C. 1983. "The Taxonomy, Reproductive Biology and Economic Potential of *Parkia* (Leguminosae: Mimosoideae) in Africa and Madagascar." *Botanical Journal of the Linnean Society* 87 (2): 135–67.

Hopkins, H. C., and F. White. 1984. "The Ecology and Chorology of *Parkia* in Africa." *Bulletin du Jardin botanique National de Belgique* 54 (1–2): 235–66.

Howes, Melanie-Jayne R., Cassandra L. Quave, Jérôme Collemare, Evangelos C. Tatsis, Danielle Twilley, Ermias Lulekal, Andrew Farlow, et al. 2020. "Molecules from Nature: Reconciling Biodiversity Conservation and Global Healthcare Imperatives for Sustainable Use of Medicinal Plants and Fungi." *Plants, People, Planet* 2: 463–81.

Hsu, Jeremy. 2017. "The Hard Truth About the Rhino Horn "Aphrodisiac" Market." *Scientific American*, April 5, 2017. www.scientificamerican.com/article/the-hard-truth-about-the-rhino-horn-aphrodisiac-market/.

Hughes, Daniel F., Mathias Behangana, Wilber Lukwago, Michele Menegon, J. Maximilian Dehling, Philipp Wagner, Colin R. Tilbury, et al. 2024. "Taxonomy of the *Rhampholeon boulengeri* Complex (Sauria: Chamaeleonidae): Five New Species from Central Africa's Albertine Rift." *Zootaxa* 5458 (4): 451–94.

Humle, T., F. Maisels, J. F. Oates, A. Plumptre, and E. A. Williamson. 2016. "*Pan troglodytes* (errata version published in 2018)." The IUCN Red List of Threatened Species 2016: e.T15933A129038584. Accessed February 2, 2025. https://dx.doi.org/10.2305/IUCN.UK.2016-2.RLTS.T15933A17964454.en.

Hunt, Nancy Rose. 2016. *A Nervous State: Violence, Remedies, and Reverie in Colonial Congo*. Durham, NC: Duke University Press.

Hunter, Frederic. 2016. *A Year at the Edge of the Jungle: A Congo Memoir: 1963–1964*. Seattle: Cune Press.

Inogwabini, Bila-Isia. 2020. *Reconciling Human Needs and Conserving Biodiversity: Large Landscapes as a New Conservation Paradigm. The Lake Tumba, Democratic Republic of Congo*. Cham, Switzerland: Springer.

Inogwabini, Bila-Isia, Matungila Bewa, Mbende Longwango, Mbenzo Abokome, and Miezi Vuvu. 2008. "The Bonobos of the Lake Tumba—Lake Maindombe Hinterland: Threats and Opportunities for Population Conservation." In *The Bonobos: Behavior, Ecology, and Conservation*, edited by Takeshi Furuichi and Jo Thompson, 273–90. New York: Springer.

Irfan, Umair. 2024. "2023 Was the Hottest Year on Record. It Also Pushed the World Over a Dangerous Line." *Vox*, February 8, 2024. www.vox.com/23969523/climate-change-cop28-paris-1-5-c-uae-2023-record-warm.

Isbell, Lynne A. 2006. "Snakes as Agents of Evolutionary Change in Primate Brains." *Journal of Human Evolution* 51 (1): 1–35.

IUCN 2021. "Issues Brief: Peatlands and Climate Change." November 2021. www.iucn.org/sites/default/files/2022-04/iucn_issues_brief_peatlands_and_climate_change_final_nov21.pdf.

IUCN SSC Antelope Specialist Group. 2016. "*Tragelaphus eurycerus* (errata version published in 2017)." The IUCN Red List of Threatened Species 2016: e.T22047A115164600. Accessed February 19, 2025. https://dx.doi.org/10.2305/IUCN.UK.2016-3.RLTS.T22047A50195617.en.

References

Izadi, Elahe. 2016. "We Saved the Alligators from Extinction—Then Moved into Their Territory." *Washington Post*, June 17, 2016. www.washingtonpost.com/news/animalia/wp/2016/06/17/we-saved-the-alligators-from-extinction-then-moved-into-their-territory/.

Jacobson, Tyler A., Jasdeep S. Kler, Michael T. Hernke, Rudolf K. Braun, Keith C. Meyer, and William E. Funk. 2019. "Direct Human Health Risks of Increased Atmospheric Carbon Dioxide." *Nature Sustainability* 2: 691–701.

Jangannathan, Prasanna, and Abel Kakuru. 2022. "Increasing Challenges, Cautious Optimism." *Nature Communications* 13: 2678.

Jeal, Tim. 2007. *Stanley: The Impossible Life of Africa's Greatest Explorer*. New Haven, CT: Yale University Press.

Jenner, Ronald, and Eivind Undheim. 2017. *Venom: The Secrets of Nature's Deadliest Weapon*. Washington, DC: Smithsonian Books.

Jin, Rui Nian, Hitoshi Inada, János Négyesi, Daisuke Ito, and Ryoichi Nagatomi. 2022. "Carbon Dioxide Effects on Daytime Sleepiness and EEG Signal: A Combinational Approach Using Classical Frequentist and Bayesian Analyses." *Indoor Air* 32: e13055.

Johnson, Christopher N., Andrew Balmford, Barry W. Brook, Jessie C. Buettel, Mauro Galetti, Lei Guangchun, and Janet M. Wilmshurst. 2017. "Biodiversity Losses and Conservation Responses in the Anthropocene." *Science (Washington)* 356 (6335): 270–75.

Johnston, Sir Harry. 1908. *George Grenfell and the Congo: A History and Description of the Congo Independent State and Adjoining Districts of Congoland*. Vol. 1. London: Hutchinson.

Jones, Benji. 2024. "Are Rainforests Doomed? Not Necessarily." *Vox*, April 4, 2024. www.vox.com/down-to-earth/24114997/amazon-rainforest-deforestation-hope.

Jones, Kate E., Nikkita G. Patel, Marc A. Levy, Adam Storeygard, Deborah Balk, John L. Gittleman, and Peter Daszak. 2008. "Global Trends in Emerging Infectious Diseases." *Nature (London)* 451: 990–94.

Jones, Nicola. 2023. "When Will Global Warming Actually Hit the Landmark 1.5 °C Limit?" *Nature (London)*, May 19, 2023. www.nature.com/articles/d41586-023-01702-w.

Jongsma, Gregory F. M., Michael F. Barej, Christopher D. Barratt, Marius Burger, Werner Conradie, Raffael Ernst, Eli Greenbaum, et al. 2018. "Diversity and Biogeography of Frogs in the Genus *Amnirana* (Anura: Ranidae) Across Sub-Saharan Africa." *Molecular Phylogenetics and Evolution* 120: 274–85.

Joris, Lieve. 1992. *Back to the Congo*. New York: Antheum.

Joy, Charles R, ed. and trans. 1951. *The Animal World of Albert Schweitzer: Jungle Insights into Reverence for Life.* Boston: Beacon Press.

Kahlenberg, Sonya M., Tammie Bettinger, Honoré K. Masumbuko, Gracianne K. Basyanirya, Simisi M. Guy, Jonathan K. Katsongo, Nadine Kocanjer, et al. 2020. "A Case Study of Improved Cook Stoves in Primate Conservation from Democratic Republic of Congo." *American Journal of Primatology* 83 (4): e23218.

Kaiser, Josef, Dagmar Haase, and Tobias Krueger. 2021. "Payments for Ecosystem Services: A Review of Definitions, the Role of Spatial Scales, and Critique." *Ecology and Society* 26 (2): 12.

Karesh, William B., Andy Dobson, James O. Lloyd-Smith, Juan Lubroth, Matthew A. Dixon, Malcolm Bennett, Stephen Aldrich, et al. 2012. "Ecology of Zoonoses: Natural and Unnatural Histories." *Lancet* 380 (9857): 1936–45.

Kariuki, Siliva N., and Thomas N. Williams. 2020. "Human Genetics and Malaria Resistance." *Human Genetics* 139: 801–11.

Karnauskas, Kristopher B., Shelly L. Miller, and Anna C. Schapiro. 2020. "Fossil Fuel Combustion Is Driving Indoor CO_2 Toward Levels Harmful to Human Cognition." *GeoHealth* 4: e2019GH000237.

Karsten, Kristopher B., Laza N. Andriamandimbiarisoa, Stanley F. Fox, and Christopher J. Raxworthy. 2008. "A Unique Life History Among Tetrapods: An Unusual Chameleon Living Mostly as an Egg." *Proceedings of the National Academy of Sciences (USA)* 105 (26): 8980–84.

Kayumba, Mathieu, Constantin Lubini, Eustache Kidikwadi, and Jean-Pierre Habari. 2015. "Etude Floristique de la Végétation de la Formation Mature du Domaine et Réserve de Bombo-Lumene (Kinshasa/RD Congo)." *International Journal of Innovation and Applied Studies* 11 (3): 716–27.

Kehinde Omifolaji, James, Alice C. Hughes, Abubakar Sadiq Ibrahim, Jinfeng Zhou, Siyuan Zhang, Emmanuel Tersea Ikyaagba, and Xiaofeng Luan. 2022. "Dissecting the Illegal Pangolin Trade in China: An Insight from Seizures Data Reports." *Nature Conservation* 45: 17–38.

Keller, Mitch. 2006. "The Scandal at the Zoo." *New York Times*, August 6, 2006. www.nytimes.com/2006/08/06/nyregion/thecity/the-scandal-at-the-zoo.html.

Kielgast, Jos, and Stefan Lötters. 2009. "Forest Weaverbird Nests Utilized by Foam-Nest Frogs (Rhacophoridae: *Chiromantis*) in Central Africa." *Salamandra* 45 (3): 170–71.

Kilvert, Nick. 2023. "Estuarine Crocodiles 'Talk' to Each Other. Scientists Are Learning What They're Saying." *ABC Science*, August 8, 2023.

References

www.abc.net.au/news/science/2023-08-09/estuarine-crocodiles-talk-to-each-other-scientists-are-learning/102672584.

Kingdon, Jonathan. 2023. *Origin Africa: A Natural History*. Princeton, NJ: Princeton University Press.

Kingdon, Jonathan, and Michael Hoffmann, eds. 2013a. *Mammals of Africa*. Vol. 5: *Carnivores, Pangolins, Equids and Rhioceroses*. London: Bloomsbury.

Kingdon, Jonathan, and Michael Hoffmann, eds. 2013b. *Mammals of Africa*. Vol. 6: *Pigs, Hippopotamuses, Chevrotain, Giraffes, Deer and Bovids*. London: Bloomsbury.

Kingsley, Mary H. 1897. *Travels in West Africa: Congo Francais, Corisco and Cameroons*. London: Macmillan.

Kirstein, Oscar D., Roy Faiman, Araya Gebreselassie, Asrat Hailu, Teshome Gebre-Michael, and Alon Warburg. 2013. "Attraction of Ethiopian Phlebotomine Sand Flies (Diptera: Psychodidae) to Light and Sugar-Yeast Mixtures (CO_2)." *Parasites and Vectors* 6: 341.

Kirwan, G. M., A. C. Kemp, J. del Hoyo, N. Collar, and P. F. D. Boesman 2021. "African Pied Hornbill (*Lophoceros fasciatus*)" (web page). CornellLab, Birds of the World. Version 2.0. September 17, 2021. Edited by B. K. Keeney, Cornell Lab of Ornithology, Ithaca, NY. https://doi.org/10.2173/bow.afphor1.02.

Kissling, W. Daniel, Wolf L. Eiserhardt, William J. Baker, Finn Borchseniusa, Thomas L. P. Couvreur, Henrik Balsleva, and Jens-Christian Svenning. 2012. "Cenozoic Imprints on the Phylogenetic Structure of Palm Species Assemblages Worldwide." *Proceedings of the National Academy of Sciences (USA)* 109 (19): 7379–84.

Klieman, Kairn A. 2003. *"The Pygmies Were Our Compass": Bantu and Batwa in the History of West Central Africa, Early Times to c. 1900 C.E.* Portsmouth, NH: Heinemann.

Kluger, Jeffrey. 2024. "An Inside Look at the Embryo Transplant That May Help Save the Northern White Rhino." *Time*, January 25, 2024. https://time.com/6588316/breakthrough-white-rhino-embryo-implantation-photos/.

Knight, Judy. 2003. "Relocated to the Roadside: Preliminary Observations on the Forest Peoples of Gabon." *African Study Monographs*, Supplement 28: 81–121.

Koch, Alexander, Chris Brierley, Mark M. Maslin, and Simon L. Lewis. 2019. "Earth Systems Impacts of the European Arrival and Great Dying in the Americas After 1492." *Quaternary Science Reviews* 207: 13–36.

Koch, Alexander, and Jed O. Kaplan. 2022. "Tropical Forest Restoration Under Future Climate Change." *Nature Climate Change* 12: 279–83.

Kolbert, Elizabeth. 2014. *The Sixth Extinction: An Unnatural History*. New York: Picador.

Kreier, Freda. 2022. "Tropical Forests Have Big Climate Benefits Beyond Carbon Storage." *Nature (London)*, April 1, 2022. www.nature.com/articles/d41586-022-00934-6#ref-CR1.

Krishnan, Sneha, Caleb Ofori-Boateng, Matthew K. Fujita, and Adam D. Leaché. 2019. "Geographic Variation in West African *Agama picticauda*: Insights from Genetics, Morphology and Ecology." *African Journal of Herpetology* 68 (1): 33–49.

Kujirakwinja, D., A. J. Plumptre, A. Twendilonge, G. Mitamba, L. Mubalama, J. D. D. Wasso, O. Kisumbu, et al. 2019. "Establishing the Itombwe Natural Reserve: Science, Participatory Consultations and Zoning." *Oryx* 53 (1): 49–57.

Kumakamba, Charles, Fabien R. Niama, Francisca Muyembe, Jean-Vivien Mombouli, Placide Mbala Kingebeni, Rock Aime Nina, Ipos Ngay Lukusa, et al. 2021. "Coronavirus Surveillance in Wildlife from Two Congo Basin Countries Detects RNA of Multiple Species Circulating in Bats and Rodents." *PLoS ONE* 16 (6): e0236971.

Kumar, Sudhir, Koichiro Tamura, and Masatoshi Nei. 1993. *MEGA: Molecular Evolutionary Genetics Analysis*, version 1.01. Pennsylvania State University, University Park.

Kumar Tripathi, Lokesh, and Tapan Kumar Nailwal. 2021. "Leishmaniasis: An Overview of Evolution, Classification, Distribution, and Historical Aspects of Parasite and its Vector." In *Pathogenesis, Treatment and Prevention of Leishmaniasis*, edited by Mukesh Samant and Satish Chandra Pandey, 1–25. London: Academic Press.

Kusamba, Chifundera, Alan Resetar, Van Wallach, Kasereka Lulengo, and Zoltán T. Nagy. 2013. "Mouthful of Snake: An African Snake-eater's (*Polemon fulvicollis graueri*) Large Typhlopid Prey." *Herpetology Notes* 6: 235–37.

Lagamma, Alisa. 2016. *Kongo: Power and Majesty*. New York: Metropolitan Museum of Art.

Langergraber, Kevin E., Kay Prüfer, Carolyn Rowney, Christopher Boesch, Catherine Crockford, Katie Fawcett, Eiji Inoue, et al. 2012. "Generation Times in Wild Chimpanzees and Gorillas Suggest Earlier Divergence Times in Great Ape and Human Evolution." *Proceedings of the National Academy of Sciences (USA)* 109 (39): 15716–21.

Laurent, R.-F. 1965. "Contribution à l'Histoire de l'Herpétologie Congolaise et Bibliographie Générale." *Académie Royale des Sciences d'Outre Mer, Classe de Sciences Naturelles et Médicales, Nouvelle Série* 16 (3): 1–55.

References

Leaché, Adam D., Jared A. Grummer, Michael Miller, Sneha Krishnan, Matthew K. Fujita, Wolfgang Böhme, Andreas Schmitz, et al. 2017. "Bayesian Inference of Species Diffusion in the West African *Agama agama* Species Group (Reptilia, Agamidae)." *Systematics and Biodiversity* 15 (3): 192–203.

Leakey, Meave G., Craig S. Feibel, Raymond L. Bernor, John M. Harris, Thure E. Cerling, Kathlyn M. Stewart, Glenn W. Storrs, et al. 1996. "Lothagam: A Record of Faunal Change in the Late Miocene of East Africa." *Journal of Vertebrate Paleontology* 16 (3): 556–70.

LeBreton, Matthew, Otto Yang, Ubald Tamoufe, Eitel Mpoudi-Ngole, Judith N. Torimiro, Cyrille F. Djoko, Jean K. Carr, et al. 2007. "Exposure to Wild Primates Among HIV-Infected Persons." *Emerging Infectious Diseases* 13 (10): 1579–82.

Ledgard, J. M. 2014. "History's Stranglehold." *New York Times*, May 1, 2014. www.nytimes.com/2014/05/04/books/review/david-van-reybroucks-congo.html.

Lee, Joshua. 2022. "Don't Call Yourself an 'American'—Stanford Initiative Is Criticized Online." *Deseret News*, December 20, 2022. www.deseret.com/u-s-world/2022/12/20/23519767/dont-call-yourself-an-american-stanford-initiative-is-criticized-online/.

Lee, Song Hee, Eui Ho Kim, Justin T. O'Neal, Gordon Dale, David J. Holthausen, James R. Bowen, Kendra M. Quicke, et al. 2021. "The Amphibian Peptide Yodha Is Virucidal for Zika and Dengue Viruses." *Scientific Reports* 11: 602.

Leisher, Craig, Nathaniel Robinson, Matthew Brown, Deo Kujirakwinja, Mauricio Castro Schmitz, Michelle Wieland, and David Wilkie. 2022. "Ranking the Direct Threats to Biodiversity in Sub-Saharan Africa." *Biodiversity and Conservation* 31: 1329–43.

Leroy, Eric M., Alain Epelboin, Vital Mondonge, Xavier Pourrut, Jean-Paul Gonzalez, Jean-Jacques Muyembe-Tamfum, and Pierre Formenty. 2009. "Human Ebola Outbreak Resulting from Direct Exposure to Fruit Bats in Luebo, Democratic Republic of Congo, 2007." *Vector-Borne and Zoonotic Diseases* 9 (6): 723–28.

Li, Guangdong, Chuanglin Fang, James E. M. Watson, Siao Sun, Wei Qi, Zhenbo Wang, Jianguo Liu. 2024. "Mixed Effectiveness of Global Protected Areas in Resisting Habitat Loss." *Nature Communications* 15: 8389.

Li, Tania Murray. 2018. "After the Land Grab: Infrastructural Violence and the 'Mafia System' in Indonesia's Oil Palm Plantation Zones." *Geoforum* 96: 328–37.

Liebowitz, Daniel, and Charles Pearson. 2005. *The Last Expedition: Stanley's Mad Journey Through the Congo*. New York: Norton.

Linder, Hans Peter. 2017. "East African Cenozoic Vegetation History." *Evolutionary Anthropology* 26 (6): 300–12.

Lóbon-Rovira, Javier, Aaron M. Bauer, Pedro Vaz Pinto, Jean-François Trape, Werner Conradie, Chifundera Kusamba, Timoteo Julio, et al. 2024. "Integrative Revision of the *Lygodactylus gutturalis* (Bocage, 1873) Complex Unveils Extensive Cryptic Diversity and Traces Its Evolutionary History." *Zoological Journal of the Linnean Society* 201 (2): 447–92.

Lóbon-Rovira, Javier, Werner Conradie, David Buckley Iglesias, Raffael Ernst, Luis Veríssimo, Ninda Baptista, and Pedro Vaz Pinto. 2021. "Between Sand, Rocks and Branches: An Integrative Taxonomic Revision of Angolan *Hemidactylus* Goldfuss, 1820, with Description of Four New Species." *Vertebrate Zoology* 71: 465–501.

Lorente-Galdos, Belen, Oscar Lao, Gerard Serra-Vidal, Gabriel Santpere, Lukas F. K. Kuderna, Lara R. Arauna, Karima Fadhlaoui-Zid, et al. 2019. "Whole-Genome Sequence Analysis of a Pan African Set of Samples Reveals Archaic Gene Flow from an Extinct Basal Population of Modern Humans into Sub-Saharan Populations." *Genome Biology* 20: 77.

Loveridge, Arthur. 1931. "On Two Amphibious Snakes of the Central African Lake Region." *Bulletin of the Antivenin Institute of America* 5 (1): 7–12.

Loveridge, Arthur. 1938. "On a Collection of Reptiles and Amphibians from Liberia." *Proceedings of the New England Zoölogical Club* 17: 49–74.

Luiselli, Luca, Godfrey C. Akani, Edem A. Eniang, and Edoardo Politano. 2007. "Comparative Ecology and Ecological Modeling of Sympatric Pythons, *Python regius* and *Python sebae*." In *Biology of the Boas and Pythons*, edited by Robert W. Henderson and Robert Powell, 88–100. Eagle Mountain, UT: Eagle Mountain Publishing.

Lupo, Karen, Alfred Jean-Paul Ndanga, and Christopher Kiahtipes. 2017. "On Late Holocene Population Interactions in the Northwestern Congo Basin: When, How, and Why Does the Ethnographic Pattern Begin?" In *Hunter-Gatherers of the Congo Basin: Cultures, Histories, and Biology of African Pygmies*, edited by Barry S. Hewlett, 59–83. London: Routledge.

Lyons, Maryinez. 2002. *The Colonial Disease: A Social History of Sleeping Sickness in Northern Zaire, 1900–1940*. Cambridge: Cambridge University Press.

Maclean, Ruth, and Caleb Kabanda. 2022. "What Do the Protectors of Congo's Peatlands Get in Return?" *New York Times*, February 21, 2022. www.nytimes.com/interactive/2022/02/21/headway/peatlands-congo-climate-change.html.

Malhi, Yadvinder, Janet Franklin, Nathalie Seddon, Martin Solan, Monica G. Turner, Christopher B. Field, and Nancy Knowlton. 2020. "Climate Change and Ecosystems: Threats, Opportunities and Solutions." *Philosophical Transactions of the Royal Society B: Biological Sciences* 375 (1794): 20190104.

References

Malvy, D., and F. Chappuis. 2011. "Sleeping Sickness." *Clinical Microbiology and Infection* 17: 986–95.

Mandavilli, Apoorva. 2022. "At Long Last, Can Malaria be Eradicated?" *New York Times*, October 4, 2022. www.nytimes.com/2022/10/04/health/malaria-vaccines.html.

Mangulu, Motinega. 2003. "MBANDAKA-COQUILHATVILLE (1883–2002): Echec d'un Plan de Developpement Exogene: Une Lecture par la Chanson Populaire." *Annales Æquatoria* 24: 179–204.

Mann, Charles C. 2018. *The Wizard and the Prophet: Two Dueling Scientists and Their Dueling Visions to Shape Tomorrow's World*. New York: Vintage Books.

Mao, Yafei, Claudia R. Catacchio, LaDeana W. Hillier, David Porubsky, Ruiyang Li, Arvis Sulovari, Jason D. Fernandes, et al. 2021. "A High-Quality Bonobo Genome Refines the Analysis of Hominid Evolution." *Nature (London)* 594: 77–81.

Marr, Lisa. 2007. *Sexually Transmitted Diseases: A Physician Tells You What You Need to Know*. Baltimore: Johns Hopkins University Press.

Marsh, Charles J., Edgar C. Turner, Benjamin Wong Blonder, Boris Bongalov, Sabine Both, Rudi S. Cruz, Dafydd M. O. Elias, et al. 2025. "Tropical Forest Clearance Impacts Biodiversity and Function, Whereas Logging Changes Structure." *Science (Washington)* 387: 171–75.

Masci, Joseph R., and Elizabeth Bass. 2018. *Ebola: Clinical Patterns, Public Health Concerns*. Boca Raton, FL: CRC Press.

Maurel, Auguste. 1992. *Le Congo de la Colonisation Belge à l'Indépendence*. Paris: L'Harmattan.

McCleary, Ryan J. R., and R. Manjunatha Kini. 2013. "Non-Enzymatic Proteins from Snake Venoms: A Gold Mine of Pharmacological Tools and Drug Leads." *Toxicon* 62: 56–74.

McDowell, Robin, and Margie Mason. 2020. "Child Labor in Palm Oil Industry Tied to Girl Scout Cookies." *AP*, December 29, 2020. https://apnews.com/article/palm-oil-forests-indonesia-scouts-83b01f2789e9489569960da63b2741c4.

McKenna, Amy, ed. 2011. *The History of Central and Eastern Africa*. The Britannica Guide to Africa. New York: Britannica Educational.

McLynn, Frank. 1993. *Hearts of Darkness: The European Exploration of Africa*. New York: Carroll and Graf.

McLynn, Frank. 2004. *Stanley: Dark Genius of African Exploration*. London: Pimlico.

Mednick, Sam. 2024. "Takeaways From AP Visit to Congo as Nation Offers Expanded Oil Drilling." *AP*, February 29, 2024. https://apnews.com/article/congo-africa-oil-gas-exploration-perenco-pollution-2f3de78c60e2efce712a49b14353f230.

Megevand, Carole. 2013. *Deforestation Trends in the Congo Basin: Reconciling Economic Growth and Forest Protection.* Washington, DC: World Bank.

Meier, Jürg, and Julian White. 1995. *Handbook of Clinical Toxicology of Animal Venoms and Poisons.* Boca Raton, FL: CRC Press.

Meiri, Shai, David G. Chapple, Krystal A. Tolley, Nicola Mitchell, Timrat Laniado, Neil Cox, Phil Bowles, et al. 2023. "Done but Not Dusted: Reflections on the First Global Reptile Assessment and Priorities for the Second." *Biological Conservation* 278: 109879.

Meijaard, Erik, Thomas M. Brooks, Kimberly M. Carlson, Eleanor M. Slade, John Garcia-Ulloa, David L. A. Gaveau, Janice Ser Huay Lee, et al. 2020. "The Environmental Impacts of Palm Oil in Context." *Nature Plants* 6: 1418–26.

Milau, Fils, Carmel Kifukieto, Claude Kachaka, Jules Aloni, and Frédéric Francis. 2016. "Contribution à l'Étude de la Faune Associée à la Décomposition du Bois (Isoptera et Haplotaxida) à Bombo-Lumene au Plateau des Batékés (RDC)." *Entomologie Faunistique* 69: 37–44.

Milman, Oliver. 2024. "Record-Breaking Increase in CO2 Levels in World's Atmosphere." *Guardian*, May 9, 2024. www.theguardian.com/environment/article/2024/may/09/carbon-dioxide-atmosphere-record.

Minton, Sherman A. Jr., and Madge Rutherford Minton. 1973. *Giant Reptiles.* New York: Charles Scribner.

Mitchell, Ian, and Samuel Pleeck. 2022. "How Much Should the World Pay for the Congo Forest's Carbon Removal?" Center for Global Development, November 2, 2022. www.cgdev.org/publication/how-much-should-world-pay-congo-forests-carbon-removal.

Mitchell, Kevin. 2014. "Rumble in the Jungle: The Night Ali Became King of the World Again." *Guardian*, October 29, 2014. www.theguardian.com/sport/2014/oct/29/rumble-in-the-jungle-muhammad-ali-george-foreman-book-extract.

Moat, Justin F., and Rachel Purdon. 2021. "How Much Carbon Is Stored in Kew's Trees?" Royal Botanic Gardens, Kew, May 25, 2021. www.kew.org/read-and-watch/kew-carbon-trees-draw-down.

Mocquard, M. F. 1906. "Description de Quelques Espèces Nouvelles de Reptiles." *Bulletin du Muséum National d'Histoire Naturelle* 12 (7): 464–65.

Moelants, T. 2010. "*Nanochromis transvestitus.*" The IUCN Red List of Threatened Species 2010: e.T182295A7852784. Accessed February 19, 2025. https://dx.doi.org/10.2305/IUCN.UK.2010-3.RLTS.T182295A7852784.en.

Montilly, Juliette. 2021. "COP26: Climate Change Transforms Mali Lake into Desert, Exiling Population." *France 24*, October 29, 2021. www.france24.com/en/

References

africa/20211029-cop26-climate-change-transforms-mali-lake-into-desert-exiling-population.

Mora, Camilo, Derek P. Tittensor, Sina Adl, Alastair G. B. Simpson, and Boris Worm. 2011. "How Many Species Are There on Earth and in the Ocean?" *PLOS Biology* 9 (8): e1001127.

Morrell, Robert. 1998. "Of Boys and Men: Masculinity and Gender in Southern African Studies." *Journal of Southern African Studies* 24 (4): 605–30.

Muiruri, Peter. 2022. "'We Relied on the Lake. Now It's Killing Us': Climate Crisis Threatens Future of Kenya's El Molo People." *Guardian*, February 1, 2022. www.theguardian.com/global-development/2022/feb/01/we-relied-on-the-lake-now-its-killing-us-climate-crisis-threatens-future-of-kenyas-el-molo-people.

Mukpo, Ashoka. 2022. "Gates Foundation Among Investors Backing Troubled DRC Palm Plantation." *Mongabay*, February 8, 2022. https://news.mongabay.com/2022/02/gates-foundation-among-investors-backing-troubled-drc-palm-plantation/.

Mullen, Gary R., and W. David Sissom. 2019. "Scorpions (Scorpiones)." In *Medical and Veterinary Entomology*, 3rd ed., edited by Gary R. Mullen and Lance A. Durden, 489–504. London: Academic Press.

Mulotwa, M., M. Louette, A. Dudu, A. Upoki, and R. A. Fuller. 2010. "Congo Peafowl Use Both Primary and Regenerating Forest in Salonga National Park, Democratic Republic of Congo." *Ostrich* 81 (1): 1–6.

Murphy, John C., and Robert W. Henderson. 1997. *Tales of Giant Snakes: A Historical Natural History of Anacondas and Pythons*. Malabar, FL: Krieger Publishing.

Murphy, John C., and Tom Crutchfield. 2019. *Giant Snakes: A Natural History*. N.p.: Book Services.

Murray, Brian. 2012. "Henry Morton Stanley and the Pygmies of 'Darkest Africa.'" *Public Domain Review*, November 29, 2012. https://publicdomainreview.org/essay/henry-morton-stanley-and-the-pygmies-of-darkest-africa.

Nachman, Michael W., Elizabeth J. Beckman, Rauri C. K. Bowie, Carla Cicero, Chris J. Conroy, Robert Dudley, Tyrone B. Hayes, et al. 2023. "Specimen Collection Is Essential for Modern Science." *PLOS Biology* 21 (11): e3002318.

Nagy, Zoltán T., Václav Gvoždík, Danny Meirte, Marcel Collet, and Olivier S. G. Pauwels. 2014. "New Data on the Morphology and Distribution of the Enigmatic Schouteden's Sun Snake, *Helophis schoutedeni* (de Witte, 1922) from the Congo Basin." *Zootaxa* 3755 (1): 096–100.

Nam Dang Vu, Hoai, and Martin Reinhardt Nielsen. 2018. "Understanding Utilitarian and Hedonic Values Determining the Demand for Rhino Horn in Vietnam." *Human Dimensions of Wildlife* 23 (5): 417–32.

Nanclares, Carolina, Jimmy Kapetshi, Fanshen Lionetto, Olimpia de la Rosa, Jean-Jacques Muyembe Tamfun, Miriam Alia, and Gary Kobinger. 2016. "Ebola Virus Disease, Democratic Republic of the Congo, 2014." *Emerging Infectious Diseases* 22 (9): 1579–86.

Narayanan, Surya, Peter Christopher, Kothandapani Raman, Nilanjan Mukherjee, Ponmudi Prabhu, Maniezhilan Lenin, Sivangnanaboopathidoss Vimalraj, et al. 2023. "A New Species of Rock-Dwelling *Hemidactylus* Goldfuss, 1820 (Squamata: Gekkonidae) from the Southern Eastern Ghats, India." *Vertebrate Zoology* 73: 499–512.

Nečas, Tadeáš, Jos Kielgast, Ikechukwu G. Chinemerem, Mark-Oliver Rödel, Matej Dolinay, and Václav Gvoždík. 2024. "The Phylogenetic Position of *Hyperolius sankuruensis* (Anura: Hyperoliidae) Reveals Biogeographical Affinity Between the Central Congo and West Africa, and Illuminates the Taxonomy of *Hyperolius concolor*." *Zoological Journal of the Linnean Society* 203 (2): zlae046.

Neme, Laurel. 2017. "Triumphant Rhino Transfer Ends in Tragic Conservator Death." *National Geographic*, June 8, 2017. www.nationalgeographic.com/animals/article/wildlife-watch-rhino-transfer-south-africa-poaching-rwanda.

Neuman, Scott. 2021. "Scientists Say They Could Bring Back Woolly Mammoths. But Maybe They Shouldn't." *NPR*, September 15, 2021. www.npr.org/2021/09/14/1036884561/dna-resurrection-jurassic-park-woolly-mammoth.

Newlands, G., and C. B. Martindale. 1980. "The Buthid Scorpion Fauna of Zimbabwe-Rhodesia with Checklists and Keys to the Genera and Species, Distribution and Medical Importance (Arachnida: Scorpiones)." *Zeitschrift für Angewandte Zoologie* 67 (1): 51–77.

Nicholson, Sharon E., Chris Funk, and Andreas H. Fink. 2018. "Rainfall Over the African Continent from the 19th Through the 21st Century." *Global and Planetary Change* 165: 114–27.

Nietzel, Michael T. 2023. "Stanford and Yale Endowment Returns Are Up, While MIT and Duke See Losses Again." *Forbes*, October 13, 2023. www.forbes.com/sites/michaeltnietzel/2023/10/13/stanford-and-yale-endowment-returns-are-up-while-mit-and-duke-see-losses-again/.

Nilsen, Ella, Matt Egan, and Chris Isidore. 2025. "Trump Signs Actions to Pull US Out of Paris Climate Agreement, Intends to Promote Fossil Fuels and Mineral

References

Mining." *CNN*, January 20, 2025. www.cnn.com/2025/01/20/climate/trump-paris-agreement-energy-orders/index.html.

Nok, Andrew J. 2009. "African Trypanosomiasis." In *Vaccines for Biodefense and Emerging and Neglected Diseases*, edited by Alan D. T. Barrett and Lawrence R. Stanberry, 1255–73. London: Elsevier.

Nolen, Stephanie. 2024. "The First Vaccine for Malaria Received Major Regulatory Approval in 2015." *New York Times*, July 5, 2024. www.nytimes.com/2024/07/05/health/malaria-vaccine-delay.html.

Nuñez, Leroy P., Kenneth L. Krysko, and Michael L. Avery. 2016. "Confirmation of Introduced *Agama picticauda* in Florida Based on Molecular Analyses." *Bulletin of the Florida Museum of Natural History* 54 (9): 138–46.

Nürk, Nicolai M., H. Peter Linder, Resnske E. Onstein, Matthew J. Larcombe, Colin E. Hughes, Laura Piñeiro Fernández, Philipp M. Schlüter, et al. 2020. "Diversification in Evolutionary Arenas—Assessment and Synthesis." *Ecology and Evolution* 10 (12): 6163–82.

Nuwer, Rachel. 2018. "War's Other Victims: Animals." *New York Times*, January 12, 2018. www.nytimes.com/2018/01/12/science/africa-war-animals-conservation.html.

Oakland Institute. 2024. "Legal Battles Escalate Between Shareholders of PHC Oil Palm Plantations in DRC." *Oakland Institute*, February 29, 2024. www.oaklandinstitute.org/legal-battles-escalate-between-shareholders-phc-oil-palm-plantations-drc.

O'Hanlon, Redmond. 1997. *Congo Journey*. London: Penguin Books.

Oliveira, Ana L., Matilde F. Viegas, Saulo L. da Silva, Andreimar M. Soares, Maria J. Ramos, and Pedro A. Fernandes. 2022. "The Chemistry of Snake Venom and Its Medicinal Potential." *Nature Reviews Chemistry* 6: 451–69.

Olivero, Jesús, John E. Fa, Miguel A. Farfán, Jerome Lewis, Barry Hewletts, Thomas Breuer, Giuseppe M. Carpanet, et al. 2016. "Distribution and Numbers of Pygmies in Central African Forests." *PLoS ONE* 11 (1): e0144499.

Olson, Sarah H., Gerard Boungal, Alain Ondzie, Trent Bushmaker, Stephanie N. Seifert, Eeva Kuisma, Dylan W. Taylor, et al. 2019. "Lek-Associated Movement of a Putative Ebolavirus Reservoir, the Hammer-Headed Fruit Bat (*Hypsignathus monstrosus*), in Northern Republic of Congo." *PLoS ONE* 14 (10): e0223139.

Ordaz-Németh, Isabel, Tenekwetche Sop, Bala Amarasekaran, Mona Bachmann, Christophe Boesch, Terry Brncic, Damien Caillaud, et al. 2021. "Range-Wide Indicators of African Great Ape Density." *American Journal of Primatology* 83 (12): e23338.

Orish, Verner, Leslie Afutu, Oladapo Ayodele, Lorena Likaj, Aleksandra Marinkovic, and Adekunle Sanyaolu. 2019. "A 4-Day Incubation Period of *Plasmodium falciparum* Infection in a Nonimmune Patient in Ghana: A Case Report." *Open Forum Infectious Diseases* 6 (1): ofy169.

Oulion, Brice, James S. Dobson, Christina N. Zdenek, Kevin Arbuckle, Callum Lister, Francisco C. P. Coimbra, Bianca Op den Brouw, et al. 2018. "Factor X Activating *Atractaspis* Snake Venoms and the Relative Coagulotoxicity Neutralising Efficacy of African Antivenoms." *Toxicology Letters* 288: 119–28.

Owen-Smith, Norman. 2021. *Only in Africa: The Ecology of Human Evolution*. Cambridge: Cambridge University Press.

Page, Thomas. 2023. "A Trial Is Underway That Could Be 'The Last Roll of the Dice' for an HIV Vaccine This Decade." *CNN*, August 27, 2023. www.cnn. com/2023/08/28/health/prepvacc-hiv-vaccine-trial-spc-scn-intl/index.html.

Paice, Edward. 2022. "By 2050, A Quarter of the World's People Will Be African – This Will Shape Our Future." *Guardian*, January 20, 2022. www.theguardian. com/global-development/2022/jan/20/by-2050-a-quarter-of-the-worlds-people-will-be-african-this-will-shape-our-future.

Pang, Liping, Jie Zhang, Xiadong Cao, Xin Wang, Jin Liang, Liang Zhang, and Liang Guo. 2020. "The Effects of Carbon Dioxide Exposure Concentrations on Human Vigilance and Sentiment in an Enclosed Workplace Environment." *Indoor Air* 31: 467–79.

Parolin, Pia, and Florian Wittmann. 2010. "Struggle in the Flood: Tree Responses to Flooding Stress in Four Tropical Floodplain Systems." *AoB PLANTS* 2010: plq003.

Patin, Etienne, Katherine J. Siddle, Guillaume Laval, Hélène Quach, Christine Harmant, Noémie Becker, Alain Froment, et al. 2014. "The Impact of Agricultural Emergence on the Genetic History of African Rainforest Hunter-Gatherers and Agriculturalists." *Nature Communications* 5: 3163.

Pauwels, Olivier S. G., André Kamdem Toham, and Victor Mamonekene. 2002. "Ethnozoology of the Dibomina (Serpentes: Colubridae: *Grayia ornata*) in the Massif du Chaillu, Gabon." *Hamadryad* 27 (1): 136–41.

Pauwels, Olivier S. G., and Marc Colyn. 2023. "On a Collection of Snakes from Bengamisa, Right Bank of the Congo River, Northeastern Democratic Republic of the Congo." *Bulletin of the Chicago Herpetological Society* 58 (6): 81–92.

Pearl, Christopher A., Misty Cervantes, Monica Chan, Uyen Ho, Rane Shoji, and Eric O. Thomas. 2000. "Evidence for a Mate-Attracting Chemosignal in the Dwarf African Clawed Frog *Hymenochirus*." *Hormones and Behavior* 38: 67–74.

Pecl, Gretta T., Miguel B. Araújo, Johann D. Bell, Julia Blanchard, Timothy C. Bonebrake, I-Ching Chen, Timothy D. Clark, et al. 2017. "Biodiversity

References

Redistribution Under Climate Change: Impacts on Ecosystems and Human Well-Being." *Science (Washington)* 355 (6322): eaai9214.

Pekar, Jonathan E., Andrew Magee, Edyth Parker, Niema Moshiri, Katherine Izhikevich, Jennifer L. Havens, Karthik Gangavarapu, et al. 2022. "The Molecular Epidemiology of Multiple Zoonotic Origins of SARS-Cov-2." *Science (Washington)* 377 (6609): 960–66.

Peltier, Elian. 2023. "Battle for Influence Rages in Heart of Wagner's Operations in Africa." *New York Times*, November 26, 2023. www.nytimes.com/2023/11/26/world/africa/wagner-russia-central-african-republic.html.

Penn, James R. 2001. *Rivers of the World: A Social, Geographical, and Environmental Sourcebook*. Santa Barbara, CA: ABC-CLIO.

Pennisi, Elizabeth. 2024. "'A Tragic Mistake': Decision to Close Duke University's Herbarium Triggers Furor." *Science*, February 16, 2024. www.science.org/content/article/tragic-mistake-decision-close-duke-university-s-herbarium-triggers-furor.

Pépin, Jacques. 2021. *The Origins of AIDS*. Rev. and updated ed. Cambridge: Cambridge University Press.

Pierce, Benjamin A. 2020. *Genetics: A Conceptual Approach*. 7th ed. Austin: Macmillan Learning.

Portillo, Frank, Edward L. Stanley, William R. Branch, Werner Conradie, Mark-Oliver Rödel, Johannes Penner, Michael F. Barej, et al. 2019. "Evolutionary History of Burrowing Asps (Squamata: Lamprophiidae: Atractaspidinae) with Emphasis on Fang Evolution and Prey Selection." *PLoS ONE* 14 (5): e0214889.

Portillo, Frank, and Eli Greenbaum. 2014. "At the Edge of a Species Boundary: A New and Relatively Young Species of *Leptopelis* (Anura: Arthroleptidae) from the Itombwe Plateau, Democratic Republic of the Congo." *Herpetologica* 70 (1): 100–119.

Portillo, Frank, William R. Branch, Werner Conradie, Mark-Oliver Rödel, Johannes Penner, Michael F. Barej, Chifundera Kusamba, et al. 2018. "Phylogeny and Biogeography of the African Burrowing Snake Subfamily Aparallactinae (Squamata: Lamprophiidae)." *Molecular Phylogenetics and Evolution* 127: 288–303.

Pothasin, Pornwiwan, Stephen G. Compton, and Prasit Wangpakapattanawong. 2014. "Riparian *Ficus* Tree Communities: The Distribution and Abundance of Riparian Fig Trees in Northern Thailand." *PLoS ONE* 9 (10): e108945.

Pough, F. Harvey, Robin M. Andrews, Martha L. Crump, Alan H. Savitzky, Kentwood D. Wells, and Matthew C. Brandley. 2016. *Herpetology*. 4th ed. Sunderland, MA: Sinauer Associates.

PoWO. 2022. *Plants of the World Online*. Facilitated by the Royal Botanic Gardens, Kew. Accessed August 18, 2022. www.plantsoftheworldonline.org/.

Prat, Yosef, Mor Taub, and Yossi Yovel. 2016. "Everyday Bat Vocalizations Contain Information About Emitter, Addressee, Context, and Behavior." *Scientific Reports* 6: 39419.

Prater, Erin. 2023. "Nearly Half of the U.S. Population Has Diabetes or Prediabetes—And Many Have No Clue. Are You Among Them?" *Fortune*, December 23, 2023. https://fortune.com/well/article/diabetes-prediabetes-obesity-half-united-states-population-insulin-wegovy-type1-type2-signs-symptoms/.

Prendini, Lorenzo. 2015. "Three New *Uroplectes* (Scorpiones: Buthidae) with Punctate Metasomal Segments from Tropical Central Africa." *American Museum Novitates* (3840): 1–32.

Prüfer, Kay, Kasper Munch, Ines Hellmann, Keiko Akagi, Jason R. Miller, Brian Walenz, Sergey Koren, et al. 2012. "The Bonobo Genome Compared with the Chimpanzee and Human Genomes." *Nature (London)* 486: 527–31.

Pruitt, Bill. 1991. *African Reptiles That Have Known Me.* N.p.: Minuteman Press.

Quammen, David. 2012. *Spillover: Animal Infections and the Next Human Pandemic.* New York: Norton.

Quammen, David. 2013. "An Exclusive Look at Bonobos: The Left Bank Ape." *National Geographic Magazine* 223 (3): 98–117.

Quammen, David. 2014. *Ebola: The Natural and Human History of a Deadly Virus.* New York: Norton.

Qureshi, Adnan I. 2016. *Ebola Virus Disease: From Origin to Outbreak.* London: Academic Press.

Rabb, George B., and Mary S. Rabb. 1963. "On the Behavior and Breeding Biology of the African Pipid Frog *Hymenochirus boettgeri*." *Zeitschrift für Tierpsychologie* 20 (2): 215–41.

Ralls, Katherine. 1978. "*Tragelaphus eurycerus*." *Mammalian Species* 111: 1–4.

Raxworthy, Christopher J., and Brian Tilston Smith. 2021. "Mining Museums for Historical DNA: Advances and Challenges in Museomics." *Trends in Ecology and Evolution* 36 (11): 1049–60.

Redding, David W., Peter M. Atkinson, Andrew A. Cunningham, Gianni Lo Iacono, Lina M. Moses, James L. N. Wood, and Kate E. Jones. 2019. "Impacts of Environmental and Socio-Economic Factors on Emergence and Epidemic Potential of Ebola in Africa." *Nature Communications* 10: 4531.

Re:wild, Synchronicity Earth, IUCN SSC Amphibian Specialist Group. 2023. *State of the World's Amphibians: The Second Global Amphibian Assessment.* Texas: Re:wild.

Richardson, Maurice. 1972. *The Fascination of Reptiles.* New York: Hill and Wang.

References

Richie, Hannah, Max Roser, and Pablo Rosado. 2020. "Energy." OurWorldInData. org, https://ourworldindata.org/energy/country/democratic-republic-of-congo.

Rico-Guevara, Alejandro, Laura Echeverri-Mallarino, and Christopher J. Clark. 2022. "Oh Snap! A Within-Wing Sonation in Black-Tailed Trainbearers." *Journal of Experimental Biology* 225 (8): jeb243219.

Ringstrom, Anna, and Julie Steenhuysen. 2024. "WHO Confirms First Case of New Mpox Strain Outside Africa as Outbreak Spreads." *Reuters*, August 15, 2024. www.reuters.com/world/europe/who-confirms-first-case-new-mpox-strain-outside-africa-outbreak-spreads-2024-08-15/.

Ripple, William J., Christopher Wolf, Thomas M. Newsome, Matthew G. Betts, Gerardo Ceballos, Franck Courchamp, Matt W. Hayward, et al. 2019. "Are We Eating the World's Megafauna to Extinction?" *Conservation Letters* 12: e12627.

Roberts, Patrick. 2022. *Jungle: How Tropical Forests Shaped World History—And Us*. Dublin: Penguin Random House UK.

Rödel, Mark-Oliver, Alan Channing, and Simon N. Stuart. 2019. "Arne Schiøtz (1932–2019): A Life for African Treefrogs and Conservation." *Alytes* 37: S1–S9.

Rohwer, Vanya G., Yasha Rohwer, and Casey B. Dillman. 2022. "Declining Growth of Natural History Collections Fails Future Generations." *PLOS Biology* 20 (4): e3001613.

Rolley, Sonia, and Emma Farge. 2025. "As Congo Army Retreats from Bukavu, Children Pick Up Guns and Get Killed." *Reuters*, February 18, 2025. www.reuters.com/world/africa/congo-army-retreat-bukavu-leads-clashes-with-allied-militias-2025-02-18/.

Rookmaaker, Kees. 2013. "Family Rhinocerotidae: Rhinoceroses." *In Mammals of Africa. Volume V. Carnivores, Pangolins, Equids and Rhinoceroses*, edited by Jonathan Kingdon and Michael Hoffmann, 444–66. London: Bloomsbury.

Roosevelt, Theodore. 1987. *African Game Trails*. Camden, South Carolina: Briar Patch. First published in 1910.

Rorison, Sean. 2008. *Congo: Democratic Republic/Republic: The Bradt Travel Guide*. Guilford, CT: Globe Pequot Press.

Rosa, Isabel M. D., Matthew J. Smith, Oliver R. Wearn, Drew Purves, and Robert M. Ewers. 2016. "The Environmental Legacy of Modern Tropical Deforestation." *Current Biology* 26: 2161–66.

Roth, Kenneth, and Ida Sawyer. 2017. "The Jig Is Up for Congo's Embattled President." *Human Rights Watch*, September 19, 2017. www.hrw.org/news/2017/09/19/jig-congos-embattled-president.

Rupp, Stephanie. 2011. *Forests of Belonging: Identities, Ethnicities, and Stereotypes in the Congo River Basin*. Seattle: University of Washington Press.

Russell, Peter J. 2010. *iGenetics: A Molecular Approach*. 3rd ed. San Francisco: Benjamin Cummings.

Sage, Rowan F. 2020. "Global Change Biology: A Primer." *Global Change Biology* 26 (1): 3–30.

Sánchez-Barreiro, Fátima, Shyam Gopalakrishnan, Jazmín Ramos-Madrigal, Michael V. Westbury, Marc de Manuel, Ashot Margaryan, Marta M. Ciucani, et al. 2021. "Historical population declines prompted significant genomic erosion in the northern and southern white rhinoceros (*Ceratotherium simum*)." *Molecular Ecology* 30 (23): 6355–69.

Saragusty, Joseph, Sebastian Diecke, Micha Drukker, Barbara Durrant, Inbar Friedrich Ben-Nun, Cesar Galli, Frank Göritz, et al. 2016. "Rewinding the Process of Mammalian Extinction." *Zoo Biology* 35 (4): 280–92.

Sautter, Giles François, and Rolad Pourtier. 2025. "Congo River." *Encyclopaedia Britannica*, January 28, 2025. www.britannica.com/place/Congo-River.

Schäfer, Marvin, Karla Neira-Salamea, Laura Sandberger-Loua, Joseph Doumbia, and Mark-Oliver Rödel. 2022. "Genus-Specific and Habitat-Dependent Plant Ingestion in West African Sabre-Toothed Frogs (Anura, Odontobatrachidae: *Odontobatrachus*)." *Herpetological Monographs* 36 (1): 49–79.

Schäfermann, Simon, Cathrin Hauk, Emmanuel Wemakor, Richard Neci, Georges Mutombo, Edward Ngah Ndze, Tambo Cletus, et al. 2020. "Substandard and Falsified Antibiotics and Medicines Against Noncommunicable Diseases in Western Cameroon and Northeastern Democratic Republic of Congo." *American Journal of Tropical Medicine and Hygiene* 103 (2): 894–908.

Schick, Susanne, Jos Kielgast, Dennis Rödder, Vincent Muchai, Marius Burger, and Stefan Lötters. 2010. "New Species of Reed Frog from the Congo Basin with Discussion of Paraphyly in Cinnamon-Belly Reed Frogs." *Zootaxa* 2501: 23–36.

Schiøtz, Arne. 1982. "On Two *Afrixalus* (Anura) from Central Zaire." *Steenstrupia* 8 (11): 261–65.

Schiøtz, Arne. 2006. "Notes on the Genus *Hyperolius* (Anura, Hyperoliidae) in Central République Démocratique du Congo." *Alytes* 24: 40–60.

Schiøtz, Arne, and Helge Volsøe. 1959. "The Gliding Flight of *Holaspis guentheri* Gray, a West-African Lacertid." *Copeia* 1959 (3): 259–60.

Schlinger, Barney A. 2023. *The Wingsnappers: Lessons from an Exuberant Tropical Bird*. New Haven, CT: Yale University Press.

Schmidt, Karl Patterson. 1919. "Contributions to the Herpetology of the Belgian Congo Based on the Collection of the American Museum Congo Expedition,

1909–1915." *Bulletin of the American Museum of Natural History* 39 (2): 385–624.

Schulz, Frederik, Chantal Abergel, and Tanja Woyke. 2022. "Giant Virus Biology and Diversity in the Era of Genome-Resolved Metagenomics." *Nature Reviews* 20: 721–36.

Schweinfurth, Georg. 1874. *The Heart of Africa: Three Years's Travels and Adventures in the Unexplored Regions of Central Africa: From 1868 to 1871.* Vol. 2. Franklin Square, NY: Harper.

Schweinfurth, G., F. Ratzel, R. W. Felkin, and G. Hartlaub, eds. 1888. *Emin Pasha in Central Africa: Being a Collection of His Letters and Journals.* London: George Philip.

Sengupta, Somini, and Lisa Friedman. 2021. "U.S. Says It Will Sharply Cut Emissions and Increase Funds to Vulnerable Countries to Fight Climate Change." *New York Times*, April 22, 2021. www.nytimes.com/live/2021/04/22/us/biden-earth-day-climate-summit/us-says-it-will-sharply-cut-emissions-and-increase-funds-to-vulnerable-countries-to-fight-climate-change.

Sepulchre, Pierre, Gilles Ramstein, Frédéric Fluteau, Mathieu Schuster, Jean-Jacques Tiercelin, and Michel Brunet. 2006. "Tectonic Uplift and Eastern Africa Aridification." *Science (Washington)* 313 (5792): 1419–23.

Shanahan, Mike, Samon So, Stephen G. Compton, and Richard Corlett. 2001. "Fig-Eating by Vertebrate Frugivores: A Global Review." *Biological Reviews of the Cambridge Philosophical Society* 76 (4): 529–72.

Shapiro, Julie Teresa, Sarah Mollerup, Randi Holm Jensen, Jill Katharina Olofsson, Nam-phuong D. Nguyen, Thomas Arn Hansen, Lasse Vinner, et al. 2021. "Metagenomic Analysis Reveals Previously Undescribed Bat Coronavirus Strains in Eswatini." *EcoHealth* 18: 421–28.

Shirley, Matthew H., Amanda N. Carr, Jennifer H. Nestler, Kent A. Vliet, and Christopher A. Brochu. 2018. "Systematic Revision of the Living African Slender-Snouted Crocodiles (*Mecistops* Gray, 1844)." *Zootaxa* 4504 (2): 151–93.

Shirley, Matthew H., Kent A. Vliet, Amanda N. Carr, and James D. Austin. 2014. "Rigorous Approaches to Species Delimitation Have Significant Implications for African Crocodilian Systematics and Conservation." *Proceedings of the Royal Society B* 281 (1776): 20132483.

Shubin, Neil. 2008. *Your Inner Fish: A Journey Into the 3.5-Billion-Year History of the Human Body.* New York: Pantheon Books.

Siliceo, Gema, Jorge Morales, Mauricio Antón, and Manuel J. Salesa. 2020. "New Fossils of Amphicyonidae (Carnivora) from the Middle Miocene (MN6) Site of Carpetana (Madrid, Spain)." *Geodiversitas* 42 (15): 223–38.

Silva, Bianca. 2017. "The Story Behind the 'Rumble in the Jungle' Fight." *Time*, January 31, 2017. https://time.com/4637842/muhammed-ali-george-foreman/.

Sodhi, Navjot S., Barry W. Brook, and Corey J. A. Bradshaw. 2007. *Tropical Conservation Biology*. Malden, MA: Blackwell Publishing.

Solovieff, Nadia, Stephen W. Hartley, Clinton T. Baldwin, Elizabeth S. Klings, Mark T. Gladwin, James G. Taylor VI, Gregory J. Kato, et al. 2011. "Ancestry of African Americans with Sickle Cell Disease." *Blood Cells, Molecules, and Diseases* 47 (1): 41–45.

Southern, Lara M., Tobias Deschner, and Simone Pika. 2021. "Lethal Coalitionary Attacks of Chimpanzees (*Pan troglodytes troglodytes*) on Gorillas (*Gorilla gorilla gorilla*) in the Wild." *Scientific Reports* 11: 14673.

Spawls, Stephen, and Bill Branch. 2020. *The Dangerous Snakes of Africa*. London: Bloomsbury Wildlife.

Spawls, Stephen, Kim Howell, Harald Hinkel, and Michele Menegon. 2018. *Field Guide to East African Reptiles*. 2nd ed. London: Bloomsbury.

Stanford, Craig B., John B. Iverson, Anders G. J. Rhodin, Peter Paul van Dijk, Russell A. Mittermeier, Gerald Kuchling, Kristin H. Berry, et al. 2020. "Turtles and Tortoises Are in Trouble." *Current Biology* 30: R721–R735.

Stanley, Henry M. 1988. *Through the Dark Continent*. 2 vols. Mineola, NY: Dover. First published in 1878.

Starin, E. D., and G. M. Burghardt. 1992. "African Rock Pythons (*Python sebae*) in the Gambia: Observations on Natural History and Interactions with Primates." *Snake* 24: 50–62.

Stearns, Jason K. 2011. *Dancing in the Glory of Monsters: The Collapse of the Congo and the Great War of Africa*. New York: PublicAffairs.

Steele, A. L., T. Wibbels, and D. A. Warner. 2018. "Revisiting the First Report of Temperature-Dependent Sex Determination in a Vertebrate, the African Redhead Agama." *Journal of Zoology (London)* 306 (1): 16–22.

Stein, Rob. 2024. "Scientists Take a Step Closer to Resurrecting the Woolly Mammoth." *NPR*, March 6, 2024. www.npr.org/sections/health-shots/2024/03/06/1235944741/resurrecting-woolly-mammoth-extinction.

Stengers, J. 1957. "Combien le Congo a-t-il Coûté à la Belgique?" *Académie Royale des Sciences Coloniales, Classe des Sciences Morales et Politiques, Memoires, Nouvelle Série* 11 (1): 1–394.

References

Stewart, Donald J., and Tyson R. Roberts. 1984. "A New Species of Dwarf Cichlid Fish with Reversed Sexual Dichromatism from Lac Maindombe, Zaïre." *Copeia* 1984 (1): 82–86.

Stevenson, Colin. 2022. *Crocodiles of the World: A Complete Guide to Alligators, Caimans, Crocodiles and Gharials*. Sydney, Australia: New Holland Publishers.

Stirn, Matt. 2021. "Fleeting Glimpses of Indonesia's Endangered Orangutans." *New York Times*, August 2, 2021. www.nytimes.com/2021/08/02/travel/sumatra-orangutan-conservation.html.

Stockmann, Roland, and Eric Ythier. 2010. *Scorpions of the World*. Verrieres-Ie-Buisson, France: N.A.P. Editions.

Strochlic, Nina. 2017. "Why Pygmies Are Dealing Weed to Survive." *National Geographic*, March 21, 2017. www.nationalgeographic.com/culture/article/democratic-republic-congo-pygmy-grow-deal-weed.

Stuart, Reginald Ray. 1975. *Kassai: The Story of Raoul de Premorel, African Trader*. Stockton, CA: Pacific Center for Western Historical Studies.

Sullivan, Will. 2023. "90-Year-Old Tortoise Fathers His First Offspring at Houston Zoo." *Smithsonian Magazine*, March 24, 2023. www.smithsonianmag.com/smart-news/90-year-old-tortoise-fathers-his-first-offspring-at-houston-zoo-180981874/.

Swaroop, S. and B. Grab. 1954. "Snakebite Mortality in the World." *Bulletin of the World Health Organization* 10 (1): 35–76.

Takemoto, Hiroyuki, Yoshi Kawamoto, and Takeshi Furuichi. 2015. "How Did Bonobos Come to Range South of the Congo River? Reconsideration of the Divergence of *Pan paniscus* from Other *Pan* Populations." *Evolutionary Anthropology* 24 (5): 170–84.

Tan, Hock Heng, and Rupeng Mong. 2013. "Scorpion Stings Presenting to an Emergency Department in Singapore with Special Reference to *Isometrus Maculatus* [sic]." *Wilderness and Environmental Medicine* 24: 42–47.

Tan, Kathrine R., and Paul M. Arguin. 2017. "Malaria Diagnosis and Treatment." In *The Travel and Tropical Medicine Manual*, 5th ed., edited by Christopher A. Sanford, Paul S. Pottinger, and Elaine C. Jong, 300–314. Edinburgh: Elsevier.

Tchinda, A. T. 2008. "*Parkia bicolor* A. Chev." In PROTA (Plant Resources of Tropical Africa/Ressources végétales de l'Afrique tropicale), edited by D. Louppe, A. A. Oteng-Amoako, and M. Brink. Wageningen, Netherlands. Accessed from https://uses.plantnet-project.org/en/Parkia_bicolor_(PROTA).

Tilbury, Colin. 2018. *Chameleons of Africa*. Frankfurt: Edition Chimaira.

Timberg, Craig, and Daniel Halperin. 2012. *Tinderbox: How the West Sparked the AIDS Epidemic and How the World Can Finally Overcome It*. New York: Penguin Press.

Timmons, Jeanne. 2023. "Extinct Tasmanian Tiger Yields RNA Secrets That Could Aid Resurrection." *New Scientist*, September 19, 2023. www.newscientist.com/article/2392879-extinct-tasmanian-tiger-yields-rna-secrets-that-could-aid-resurrection/.

Tingley, Kim. 2024. "Nature's 'Swiss Army Knife': What Can We Learn from Venom?" *New York Times*, November 13, 2024. www.nytimes.com/2024/11/13/magazine/venom-animals-drugs-ozempic.html.

Tkach, Andrew, and Columbia Broadcasting System (producers). 2001. *Sleeping Sickness*. Columbia Broadcasting System, accessed from https://video.alexanderstreet.com/watch/sleeping-sickness.

Tocheri, Matthew W., René Dommain, Shannon C. McFarlin, Scott E. Burnett, D. Troy Case, Caley M. Orr, Neil T. Roach, et al. 2016. "The Evolutionary Origin and Population History of the Grauer Gorilla." *Yearbook of Physical Anthropology* 159 (S61): S4–S18.

Tolley, Krystal A., and Anthony Herrel, eds. 2013. *The Biology of Chameleons*. Berkeley: University of California Press.

Tolley, Krystal A., Graham J. Alexander, William R. Branch, Philip Bowles, and Bryan Maritz. 2016. "Conservation Status and Threats for African Reptiles." *Biological Conservation* 204 (Part A): 63–71.

Trape, Jean-François, and Youssouph Mané. 2006. "Le genre *Dasypeltis* Wagler (Serpentes: Colubridae) en Afrique de l'Ouest: Description de Trois Espèces et d'une Sous-espèce Nouvelles." *Bulletin de la Société Herpétologique de France* 119: 27–56.

Turnbull, Colin M. 1968. *The Forest People*. New York: Simon and Schuster.

Turner, D. A. and G. M. Kirwan 2020. "Great Blue Turaco (*Corythaeola cristata*)" (web page). CornellLab, Birds of the World. Version 1.0. March 4, 2020. Edited by J. del Hoyo, A. Elliott, J. Sargatal, D. A. Christie, and E. de Juana. Cornell Lab of Ornithology, Ithaca, NY. https://doi.org/10.2173/bow.grbtur1.01.

Turtle Taxonomy Working Group [Anders G. J. Rhodin, John B. Iverson, Roger Bour, Uwe Fritz, Arthur Georges, H. Bradley Shaffer, and Peter Paul van Dijk]. 2021. "Turtles of the World: Annotated Checklist and Atlas of Taxonomy, Synonymy, Distribution, and Conservation Status." 9th ed. *Chelonian Research Monographs* 8: 1–472.

Uetz, P., P. Freed, R. Aguilar, F. Reyes, J. Kudera, and J. Hošek, eds. 2025. The Reptile Database, accessed February 15, 2025. www.reptile-database.org.

References

Umutesi, Marie Béatrice. 2004. *Surviving the Slaughter: The Ordeal of a Rwandan Refugee in Zaire*. Madison, WI: University of Wisconsin Press.

United Nations. 2024. "World News in Brief: Security Council Condemns DR Congo Attacks, Cholera Testing Breakthrough, 'My Health, My Right' Campaign." *United Nations News*, April 4, 2024. https://news.un.org/en/story/2024/04/1148301.

Urban, Mark C. 2024. "Climate Change Extinctions." *Science (Washington)* 386 (6726): 1123–28.

Van Cakenberghe, Victor, Guy-Crispin Gembu Tungaluna, Prescott Musabaakawa, Ernest Seamark, and Erik Verheyen. 2017. "The Bats of the Congo and of Rwanda and Burundi Revisited (Mammalia: Chiroptera)." *European Journal of Taxonomy* (382): 1–327.

van Ittersum, Martin K., Lenny G. J. van Bussel, Joost Wolf, Patricio Grassini, Justin van Wart, Nicolas Guilpart, et al. 2016. "Can Sub-Saharan Africa Feed Itself?" *Proceedings of the National Academy of Sciences (USA)* 113 (52): 14964–69.

Van Krunkelsven, Ellen, Inogwabini Bila-Isia, and Dirk Draulans. 2000. "A Survey of Bonobos and Other Large Mammals in the Salonga National Park, Democratic Republic of Congo." *Oryx* 34 (3): 180–87.

Van Reybrouck, David. 2014. *Congo: The Epic History of a People*. New York: Ecco.

Van Schuylenbergh, Patricia. 2019. *Fauna Sauvage et Colonisation: Une Histoire de Destruction et de Protection de la Nature Congolaise (1885–1960)*. Brussels: Peter Lang.

van Seventer, Jean Maguire, and Natasha S. Hochberg. 2017. "Principles of Infectious Diseases: Transmission, Diagnosis, Prevention, and Control." In *International Encyclopedia of Public Health*, 2nd ed., edited by Stella R. Quah, 22–39. Amsterdam: Academic Press.

Vande weghe, Jean P., and Gaël R. Vande weghe. 2018. *Salonga: Au Cœur de la Grande Forêt Congolaise*. Kampenhout, Belgium: Artoos group.

Vanhooydonck, Bieke, Greet Meulepas, Anthony Herrel, Renaud Boistel, Paul Tafforeau, Vincent Fernandez, and Peter Aerts. 2009. "Ecomorphological Analysis of Aerial Performance in a Non-Specialized Lacertid Lizard, *Holaspis guentheri*." *Journal of Experimental Biology* 212 (15): 2475–82.

Vansina, Jan. 1990. *Paths in the Rainforests: Toward a History of Political Tradition in Equatorial Africa*. Madison: University of Wisconsin Press.

Vargas, Pablo, and Rafael Zardoya, eds. 2014. *The Tree of Life*. Sunderland, MA: Sinauer Associates.

Verdu, Paul. 2016. "Quick Guide: African Pygmies." *Current Biology Magazine* 26: R12–R14.

Verdu, Paul. 2017. "Population Genetics of Central African Pygmies and Non-Pygmies." In *Hunter-Gatherers of the Congo Basin: Cultures, Histories, and Biology of African Pygmies*, edited by Barry S. Hewlett, 31–58. London: Routledge.

Vernon, Géraldine, and Bruce J. Winney. 2000. "Phylogenetic Relationships Within the Turacos (Musophagidae)." *Ibis* 142 (3): 446–56.

Vidal, John. 2018. "The 100 Million City: Is 21st Century Urbanisation Out of Control?" *Guardian*, March 19, 2018. www.theguardian.com/cities/2018/mar/19/urban-explosion-kinshasa-el-alto-growth-mexico-city-bangalore-lagos.

Vijay, Varsha, Stuart L. Pimm, Clinton N. Jenkins, and Sharon J. Smith. 2016. "The Impacts of Oil Palm on Recent Deforestation and Biodiversity Loss." *PLoS ONE* 11 (7): e0159668.

Vince, Gaia. 2022. "Is the World Ready for Mass Migration Due to Climate Change?" *BBC*, November 17, 2022. www.bbc.com/future/article/20221117-how-borders-might-change-to-cope-with-climate-migration.

Wagner, David L. 2020. "Insect Declines in the Anthropocene." *Annual Review of Entomology* 65: 457–80.

Wagner, Philipp, Esther Townsend, Michael Barej, Dennis Rödder, and Stephen Spawls. 2009. "First Record of Human Envenomation by *Atractaspis congica* Peters, 1877 (Squamata: Atractaspididae)." *Toxicon* 54 (3): 368–72.

Wallace-Wells, David. 2022. "The New World: Envisioning Life After Climate Change." *New York Times*, October 26, 2022. www.nytimes.com/interactive/2022/10/26/magazine/visualization-climate-change-future.html.

Wallace-Wells, David. 2024. "Just How Many People Will Die from Climate Change?" *New York Times*, February 22, 2024. www.nytimes.com/2024/02/22/opinion/environment/climate-change-death-toll.html.

Walsh, Declan. 2023. "The World Is Becoming More African." *New York Times*, October 28, 2023. www.nytimes.com/interactive/2023/10/28/world/africa/africa-youth-population.html.

Wang, Gang, Zingtan Zhang, Edward A. Herre, Doyle McKey, Carlos A. Machado, Wen-Bin Yu, Charles H. Cannon, et al. 2021. "Genomic Evidence of Prevalent Hybridization Throughout the Evolutionary History of the Fig-Wasp Pollination Mutualism." *Nature Communications* 12: 718.

Warrell, David A. 1995. "Clinical Toxicology of Snakebite in Africa and the Middle East/Arabian Peninsula." In *Handbook of Clinical Toxicology of Animal Venoms and Poisons*, edited by Jürg Meier and Julian White, 433–92. Boca Raton, FL: CRC Press.

Weinstein, Scott A., David A. Warrell, Karim Daoues, and Nicolas Vidal. 2021. "The First Reported Snakebite by an African Snake-Eater, *Polemon* spp.

References

(Atractaspididae, Aparallactinae); Local Envenoming by Reinhardt's Snake-Eater, *Polemon acanthias* (Reinhardt, 1860)." *Toxicon* 200: 92–95.

Weinstein, Scott A., James J. Schmidt, and Leonard A. Smith. 1991. "Lethal Toxins and Cross-Neutralization of Venoms from the African Water Cobras, *Boulengerina annulata annulata* and *Boulengerina christyi*." *Toxicon* 29 (11): 1315–27.

Werdelin, Lars, and Stéphane Peigné. 2010. "Carnivora." In *Cenozoic Mammals of Africa*, edited by Lars Werdelin and William Joseph Sanders, 603–57. Berkeley: University of California Press.

Weterings, Robbie, and Kai C. Vetter. 2018. "Invasive House Geckos (*Hemidactylus* spp.): Their Current, Potential and Future Distribution." *Current Zoology* 64 (5): 559–73.

Wiens, John J., and Kristen E. Saban. 2025. "Questioning the Sixth Mass Extinction." *Trends in Ecology and Evolution* 40 (4): 375–84.

Wilcox, Christie. 2016. *Venomous: How Earth's Deadliest Creatures Mastered Biochemistry*. New York: Farrar, Straus and Giroux.

Williams, Brooke A., Hawthorne L. Beyer, Matthew E. Fagan, Robin L. Chazdon, Marina Schmoeller, Starry Sprenkle-Hyppolite, Bronson W. Griscom, et al. 2024. "Global Potential for Natural Regeneration in Deforested Tropical Regions." *Nature (London)* 636: 131–37.

Williams, Catherine, Alexander Kirby, Arsalan Marghoub, Loïc Kéver, Sonya Ostashevskaya-Gohstand, Sergio Bertazzo, Mehran Moazen et al. 2022. "A Review of the Osteoderms of Lizards (Reptilia: Squamata)." *Biological Reviews* 97 (1): 1–19.

Wilson, Van G. 2022. *Viruses: Intimate Invaders*. Cham, Switzerland: Springer Nature.

Winsor, Morgan. 2018. "Black Rhinos Return to Chad, Nearly 50 Years After Local Extinction." *ABC News*, May 3, 2018. https://abcnews.go.com/International/News/black-rhinos-return-chad-50-years-local-extinction/story?id=54908456.

Woods, Vanessa. 2010. *Bonobo Handshake: A Memoir of Love and Adventure in the Congo*. New York: Gotham Books.

Wrangham, Richard W. 1993. "The Evolution of Sexuality in Chimpanzees and Bonobos." *Human Nature* 4 (1): 47–79.

Wrangham, Richard W. 2023. "Hypotheses for the Evolution of Bonobos: Self-Domestication and Ecological Adaptation." In *Bonobos and People at Wamba: 50 Years of Research*, edited by Takeshi Furuichi, Gen'ichi Idani, Daiji Kimura, Hiroshi Ihobe, and Chie Hashimoto, 521–44. Singapore: Springer.

Wrong, Michela. 2002. *In the Footsteps of Mr. Kurtz: Living on the Brink of Disaster in Mobutu's Congo*. New York: Perennial.

Yang, John, Lorna Baldwin, and Harry Zahn. 2024. "Can Science Save the Northern White Rhino from Extinction and Even Bring Back the Dodo?" *PBS News*, March 2, 2024. www.pbs.org/newshour/show/can-these-scientific-breakthroughs-save-the-northern-white-rhino-from-extinction.

Yoshida, Junki, Atsushi Hori, Yoshitsugu Kobayashi, Michael J. Ryan, Yuji Takakuwa, and Yoshikazu Hasegawa. 2021. "A New Goniopholidid from the Upper Jurassic Morrison Formation, USA: Novel Insight into Aquatic Adaptation Toward Modern Crocodylians." *Royal Society Open Science* 8 (12): 210320.

Zachos, James, Mark Pagani, Lisa Sloan, Ellen Thomas, and Katharina Billups. 2001. "Trends, Rhythms, and Aberrations in Global Climate 65 Ma to Present." *Science (Washington)* 292 (5517): 686–93.

Zanon, Sibélia. 2020. "Amazon Initiative Pays Farmers and Ranchers to Keep the Forest Standing." *Mongabay*, November 24, 2020. https://news.mongabay.com/2020/11/amazon-initiative-pays-farmers-and-ranchers-to-keep-the-forest-standing/.

Zarin, Daniel. 2022. "The World's Peatlands Are Climate Bombs Waiting to Detonate." *New York Times*, November 5, 2022. www.nytimes.com/2022/11/05/opinion/climate-warming-peatlands.html.

Zhang, Huarong, Gary Ades, Mark P. Miller, Feng Yang, Kwok-wai Lai, and Gunter A. Fischer. 2020. "Genetic Identification of African Pangolins and their Origin in Illegal Trade." *Global Ecology and Conservation* 23: e01119.

Zhou, Xuehong, Qiang Wang, Wei Zhang, Yu Jin, Zhen Wang, Zheng Chai, Zhiqiang Zhou, et al. 2018. "Elephant Poaching and the Ivory Trade: The Impact of Demand Reduction and Enforcement Efforts by China from 2005–2017." *Global Ecology and Conservation* 16: e00486.

Zimmer, Carl. 2022. "What Makes Your Brain Different from a Neanderthal's?" *New York Times*, September 8, 2022. www.nytimes.com/2022/09/08/science/human-brain-neanderthal-gene.html?smid=nytcore-ios-share&referringSource=articleShare.

Zoer, Roland P. 2012. "The Bush Meat and Conservation Status of the African Dwarf Crocodile *Osteolaemus tetraspis*." MSc thesis, University of Pretoria, South Africa.

ACE. *See* angiotensin-converting enzyme
acquired immunodeficiency syndrome
(AIDS), 157
ACS. *See* acute coronary syndrome
acute coronary syndrome (ACS), 110
Adeosun, Walé, 175
admixture (genetic), 94
aestivation, 166–67
AFDL. *See* Alliance of Democratic Forces for
the Liberation of Congo-Zaire
Africa, 11; chameleons in, 166–68; diseases
shaping history of, 62–64, 80, 157;
gazelle fat used in, 164; and globalization of disease, 200–202; mambas in,
108; man-eating snakes in, 187–90;
militaries of, 7–8; in Miocene, 48–49;
pangolins in, 107–8; pet trade in,
17; quintessential village hallmarks,
21; turtle species in, 180–82; writing
about venomous snakes of, 87. *See also*
Central Africa; Congo; Congo River,
expedition on; Democratic Republic
of the Congo (DRC); Republic of
the Congo; West Africa; *and various
country names*
African armored cricket, 61
African house gecko (*Hemidactylus
mabouia*), 22
African lion (*Panthera leo*), 11, 22, 24, 141,
164, 216n18
African locust-bean tree (*Parkia bicolor*), 97
African oil palm (*Elaeis guineensis*), 108, 117,
173–76, 199
African Parks (conservation organization), 13
African plated lizards (*Gerrhosaurus*), 69–70
African water snake (*Grayia ornata*), 39,
97–98
Afrixalus (frog genus). *See* spiny reed frogs
Afrixalus equatorialis. See spiny reed frogs
Afropavo congensis. See Congo peafowl
Agama picticauda. See West African rainbow
lizard

Agta, Indigenous people, 189–90. *See also*
Negritos, Agta
AIDS. *See* acquired immunodeficiency
syndrome
Albertine Rift Mountains, 4, 51
Alchornea plant, 4
Ali, Muhammad, 9
Alliance of Democratic Forces for the Libera-
tion of Congo-Zaire (AFDL), 177
Alligator mississippiensis. See American
alligator
Amazon, x, 79, 199–200
American alligator (*Alligator mississippien-
sis*), 139
American crocodile (*Crocodylus acutus*), 138
American Museum Congo Expedition, 134
American Museum of Natural History,
219n65
Amiet, Jean-Louis, 65
amphibians, 46, 67, 100, 138, 147, 198;
discovering new species of, 64–65; in
Florida, USA, 17; mutilation of, 196;
pushing to extinction, xi–xii; in surveys
of national parks, 122–23. *See also* forest
toad; frogs; *various species names*
amphicyonids. *See* bear-dogs
angiotensin-converting enzyme (ACE), 110
Angola, 7–8, 66, 215n78, 220n24
Angolan green snake (*Philothamnus
angolensis*), 70
Anopheles mosquitoes, 83
antelopes, 24, 102, 140, 163. *See also* bongo
(*Tragelaphus eurycerus*); Uganda Kob
(*Kobus kob*)
"anthropized" environments, taking
advantage of, 79
Anthropocene, 201
anthropologists, 91, 93
Antoine (mayor), 143–44. *See also* Watsi
Kengo
Aparallactinae (snake subfamily), 98
Aparallactus (snake genus), 98

Index

Index

Epulu (village), 166, 179, 193
Equatorial Guinea, 166, 217n24
ethnographers, 91
Eudorcas thomsonii. See Thomson's gazelle
Eukarya, 156
Euphorbiaceae (plant family), 4, 212n4
Europe, 7–9, 53, 63, 83, 86, 88, 91, 119–20,
124, 158, 174, 184, 187, 200–201, 214n55
EVEs. *See* endogenous viral elements
evolution, 4–5, 51, 56, 153, 157, 195;
evolutionary lineages, 11, 23, 89,
137; evolutionary biology, 48, 109;
"evolutionary significant unit," 90;
evolutionary history, 90, 140, 155,
215n77; evolutionary relationships, 98,
195, 205; adaptive evolution, 106; of
primates, 109; coevolutionary arms race,
110; coevolutionary history, 222n54
exenatide, drug, 110
extinction, xi–xii, 11, 48, 60, 100, 136,
139, 169, 174, 181–82, 194–98, 200,
214n60; de-extinction, 195; "escalator to
extinction," 198; "extinction cascade,"
200; mass extinction, 66, 136, 196

fantastic reed frog (*Hyperolius phantasticus*),
104
fasciculin, toxin, 108
Feinstein, Dianne, 160
fens. *See* peatlands
Ficus (tree genus), 81
Fimi River, 42
fires, 3, 26, 29, 41, 43, 49, 54, 58, 67, 82–83,
103, 145, 151, 162, 168, 173, 175, 199,
201
firewood, xi, 56, 66, 104, 125, 144, 147, 188,
200, 202
First Congo War, 12, 177
fish, 6, 17, 29, 39, 71, 73, 76, 81, 87, 94,
100, 102, 126, 129, 138, 141, 147, 149,
153, 160, 181, 187; catfish, 148, 150;
corralling, 73; fish tank, 181
fishermen, 37–40, 52, 75, 82, 87–89, 94–95,
127–28, 147, 149, 159, 164, 180–81
fishing eagles (*Icthyophaga vocifer*), 127
fishing frogs (*Aubria*), 128
fishing villages, 35–37, 68, 74–75, 127, 129,
141, 148–49, 165, 169, 172
fishpond, 44–45

flapshell turtles, 180–82
flying foxes, 78. *See also* bats
foam-nest frogs (*Chiromantis rufescens*),
122–23
Force Publique, 74–75, 124
Foreman, George, 9
forest people. *See* Batwa, people; pygmies
forest toads (*Sclerophrys*), 77, 98, 128, 148
forest treefrogs (*Leptopelis*), 59–60, 104, 128,
131, 148, 160
forests: bonobos hiding in, 46–47; creating
negative "ripple effect" in, 78–79; gallery
forest, 19–20, 25; people of, 91–94, 97,
99, 102–3, 180, 189; tropical forests, x,
xii, 53, 169, 181, 199–201
fossils, 48, 51, 136, 155, 190, 217n9
fossil fuels, xi, 117, 199
France, xii, 157
Franquet's epauletted fruit bat (*Epomops
franqueti*), 79
frogs, 73, 132, 144–45, 160, 165, 166; Batwa
people finding, 115; calling for mates,
131; ideal conditions for, 127–28;
discovering new species of, 64–66,
101–6; locating in leaf litter, 76–77;
Mpote-Emange samples, 98, 117; in
national parks, 122–23; patterns found
in, 148; rare occurrence among, 97; skin
glands, 86; specimens found near Lotulo,
150–53; poison in, 212n7. *See also*
various species
Fromont, Cécile, 212n19
fruit bats, 77–79, 81
funding, challenge of, 194, 209
fungi, xi, 27, 199
Furcifer labordi. See Labord's chameleon

GAA. *See* Global Amphibian Assessment
Gabon, 25, 39, 51, 87, 93, 140, 163, 166, 181,
188, 193, 217n25
Gabon turtle (*Pelusios gabonensis*), 150
Gabonese clawed frog (*Xenopus mellotropi-
calis*), 25
Galloway, Jerry, 94, 224n25
Garamba National Park, 12–13
Gastropholis (lizard genus). *See* keel-bellied
lizards
Gates, Bill, 198
Gavialis gangeticus. See gharial

Index

reticulated python (*Malayopython reticulatus*), 163, 189
Rhamnophis aethiopissa. See large-eyed green tree snake
Rhampholeon boulengeri. See Boulenger's pygmy chameleon
rhinoceros viper (*Bitis nasicornis*), 147
ribonucleic acid (RNA), 155–58, 195
Richardson, Maurice, 164
RNA. *See* ribonucleic acid
rocket frogs (*Ptychadena*), 25, 44, 60, 95–96, 98
Roosevelt, Theodore, 11
Rorison, Sean, 220n19
Rousettus aegyptiacus. See Egyptian fruit bat
Ruki River, 119, 123, 125–26, 173
rumba music, 15, 29, 120, 131–33
Russia, 202
Rwanda, 5, 10, 16, 80, 84, 120, 164, 177, 186, 202

sabre-toothed frogs. *See Odontobatrachus* (frog genus)
Salonga National Park, 46, 121–23, 148
Salonga River, 121, 128, 141, 150, 163; rejoining, 159–62
saltwater crocodiles (*Crocodylus porosus*), 137
sand snakes (*Psammophis lineatus*), 56–57
Sarcosuchus imperator (extinct crocodile relative), 136
SARS-CoV-2, 158, 221n40
saw-scaled vipers (*Echis*), 58
Schiøtz, Arne, 101–2, 122, 152
Schouteden's sun snake (*Helophis schoutedeni*), 98
Schweitzer, Albert, 188–89
Sclerophrys (toad genus), 25, 44, 77, 98, 104, 128, 148, 150. *See also* forest toads
Sclerophrys camerunensis. See Cameroon toad
scorpions, 179; identifying, 130; learning about survival from, 66–67; venom of, 66–67, 130, 219n68
SDT. *See* Snake Detection Theory
Second Congo War, 10
secondary succession, 201
slaves, 8, 47, 94, 200; slave ships, 22, slaver, 188

sleeping sickness, 164–65. *See also* Human African Trypanosomiasis (HAT)
small "founder" population, 49
Smutsia gigantea. See giant pangolin
Snake Detection Theory (SDT), 109–10
snakebites, 5, 14, 39, 87, 98, 112, 225n79; treating, 108–11
snakes: arboreal snakes, 81, 108; bizarre adaptations of, 56–57; egg-eating snakes, 57–58; finding, 97–98; finding water cobras, 87–90; forest snakes, 39; identifying first snake of expedition, 36–41; man-eating snakes, 187–90; nocturnal snakes, 112; and parental care, 138; purchasing for protein, 38–39; researching venoms of, 109–11; rock pythons, 163–64; superstition regarding, 39; tree snakes, 112, 144, 150, 159; venomous snakes, 20, 87, 109, 111; water snakes, 39, 41, 73, 97, 128, 180. *See also* various species
Soeurs de l'Immaculée Conception (Sisters of the Immaculate Conception), 68
softshell turtles, 56, 180–82
soil, 29, 43, 76, 116, 175, 199; acidity in Central Africa, 93
South Africa, 11, 13, 73, 234n71
South America, 86, 136
South American Cascabel rattlesnake (*Crotalus durissus*), 110–11
South American river turtle (*Podocnemis expansa*), 227n13
South Sudan, 11–12, 44, 164
southern white rhino (*Ceratotherium simum simum*), 11–13
Sparrowhawk Reception Center. *See* Centre d'Accueil Epervier
species, 11, 23; cross-species emotional empathy, 139; cryptic species, 23, 112, 140, 168, 231n12, 231n16; endangered species, 51–52, 122, 139, 174–75; endemic species, 42; invasive species, xi, 139; keystone species, 81; new species (to science), xi–xii, 3, 23, 25, 65–67, 87–89, 98, 102, 118, 123, 128, 131, 148, 150, 165, 192–93, 196, 215n78, 230n37, 231n12; species formation, 48; taxonomic work and species, 23
Species Survival Commission (SSC), 139

Index

Wagner Group, 202
Wandege, crewmember. *See* Congo River, expedition on; Democratic Republic of the Congo (DRC)
war, 18, 28, 51, 53, 55, 62, 119, 177–78, 200, 202, 218n30; of Central Africa, 3; in kingdom of Kongo, 7–8; knives of, 30; shadow wars, 202; World War II, 50, 99, 101, 193. *See also* First Congo War; Second Congo War
water chevrotain (*Hyemoschus aquaticus*), 141
water cobras, finding, 87–90
water snakes (*Grayia*), 128
Watsi Kengo, 121–22, 141–43, 145, 147–50, 154, 160, 162–63, 178
Wegovy, 110
Wene, Ntinu, 7
West Africa, 22, 51, 62, 90, 97–98, 140–41, 158
West African crocodile (*Crocodylus suchus*), 140
West African powdered tree snake (*Toxicodryas pulverulenta*), 112
West African rainbow lizard (*Agama picticauda*), 17, 186
West African slender-snouted crocodile (*Mecistops cataphractus*), 138–41
white-bellied pangolin (*Phataginus tricuspis*), 107

white-lipped frog (*Hylarana albolabris*), 25, 60, 77, 98, 104, 128, 150, 176, 231n12
WHO. *See* World Health Organization
woolly mammoth (*Mammuthus primigenius*), 214n60
World Heritage Convention, 226n18
World Health Organization (WHO), 80, 111, 162, 212n5
World Heritage in Danger, 122–23
Wrangham, Richard, 51

Xenopus (frog genus), 98, 128, 215n77. *See also* clawed frogs
Xenopus mellotropicalis. See Gabonese clawed frog

Yenge River, 121, 149, 151, 153, 159

Zaire ebolavirus, outbreak of. *See* Ebola, virus
Zaire snake-eater (*Polemon robustus*), 99, 165
Zaïre, naming, 9–10
Zaireanisation, 120
Zepbound, 110
Zoological Record, database, 122
"zoonotic spillover" events, 79
zoonotic virus, 80